SOFT TISSUE PAIN
AND DISABILITY

Reprinted under authority
of Presidential decree No. 285
(As Amended by P.D. 400 & 1203)

by

C & E PUBLISHING CO.
1353 Dapitan St., Sampaloc, Manila

Tel. 711-65-33
732-1329

Printed by C & E Printers
1353 Dapitan St., Sampaloc, Manila

EDITION 2

SOFT TISSUE PAIN AND DISABILITY

RENE CAILLIET, M.D.

Chairman and Professor
Department of Rehabilitative Medicine
University of Southern California
School of Medicine
Los Angeles, California

Illustrations by R. Cailliet, M.D.

 F. A. DAVIS COMPANY • Philadelphia

Also by Rene Cailliet:

Foot and Ankle Pain
Hand Pain and Impairment
Knee Pain and Disability
Low Back Pain Syndrome
Neck and Arm Pain
Scoliosis
The Shoulder in Hemiplegia
Shoulder Pain

Copyright © 1988 by F. A. Davis Company
Second printing 1989

Library of Congress Cataloging-in-Publication Data

Cailliet, Rene.
  Soft tissue pain and disability.

  Includes bibliographies and index.
  1. Pain. I. Title. [DNLM: 1. Collagen Diseases. 2. Connective Tissue.
3. Musculoskeletal System. 4. Pain. WD 375 C134s]
RB127.C34 1988   616'.0472   88-3806
ISBN 0-8036-1631-7

Preface to
Second Edition

As the years passed since the first edition of *Soft Tissue Pain and Disability*, there have been many concepts and clinical observations that have been constantly revised. Many painful and disabling musculoskeletal conditions have been "discovered," recognized, or better understood.

Soft tissue as the site of pain and disability remains the major contributing consideration. With the increase in sports activities, physical conditioning, exercises for cardiovascular conditioning, and the universal attempt to beneficially influence the effects of aging after age 35, soft tissues have been subjected to stresses.

Overuse is frequently mentioned in medical and lay literature. This term is used with increasing frequency by the news media. The overuse syndrome, a new revision of an old concept, applies to the soft tissues: articular, muscular, ligamentous, and bony. The importance of these tissues has too long been overlooked or minimized.

In this new edition of *Soft Tissue Pain and Disability*, many chapters have been completely rewritten. The chapter on the low back, a subject so long an enigma to the patient and the physician, has been revised and updated to include recent discoveries. The material on the cervical spine, with its increasing frequency of injury from automobile accidents and sports injuries, has been expanded. The knee, especially subject to sports injuries, has undergone greater evaluation as to mechanism of injury, diagnostic studies, and therapeutic approaches. The lower leg, including the ankle, has also been increasingly subjected to trauma and thus its discussion has been amplified in this volume.

Also, the concept of pain is presented in greater depth. The acceptance of pain as a disease entity rather than merely a symptom is considered, as is the evolution of acute pain into chronic pain. The theories of pain are revised almost daily and new journals regarding pain are constantly emerging. Soft tissue is increasingly being recognized as the site of nociceptive pain, and health care professionals are becoming more conversant with the numerous diagnostic labels and current therapies for these conditions.

All of these factors influenced me to review, revise, and rewrite a new edition of *Soft Tissue Pain and Disability*.

RENE CAILLIET, M.D.

Preface to First Edition

"The phenomenon of pain belongs to that borderline between the body and the soul about which it is so delightful to speculate from the comfort of an arm chair, but which offers such formidable obstacles to scientific inquiry." So wrote Killgren in the introduction of his classic treatise on deep pain sensibilities.*

A large percentage of complaints that cause a patient to seek medical care is that of pain and disability of a moving part of the body. Medical practitioners of all disciplines glibly discuss pain and impairment on the basis of *soft tissue* injury, stress, sprain, or inflammation, yet soft tissues are not well defined in medical dictionaries, are not named in Nomina Anatomica, and are not so designated in most anatomic or orthopedic texts. Unfortunately the soft tissues are not considered an organ system and, thus, are not on the curriculum of medical schools. Except for occasional symposia or postgraduate courses on musculoskeletal syndromes, the practicing physician receives no education in this area. This results in less knowledge and less interest in the scope of soft tissue damage. Therefore, it is the patient who suffers from this insufficient knowledge and concern— both in pain and in adequate evaluation and treatment.

A basic knowledge of soft tissue and normal functional anatomy of the part(s) involved in pain and disability is mandatory for a meaningful evaluation of the patient. Tissues capable of evoking pain and limitation must be recognized and faulty mechanism of joints must be considered. More meaningful therapy results from this knowledge and many painful, disabling conditions are prevented or decreased in their severity and dura-

---

*Killgren, J.H.: *On the distribution of pain arising from deep somatic structures with charts of segmental pain areas.* Clin. Sci. 4:36, 1939.

tion. Many complications requiring more drastic care are prevented. Even psychologic correction can result only if the defective physiologic abnormality is recognized and understood.

The purpose of this text is to evaluate and clarify the basis of many of the painful musculoskeletal and neuromuscular conditions that beset man and perplex his physician. It is hoped that this book, along with being informative, will also stimulate more interest in soft tissue abnormalities.

RENE CAILLIET, M.D.

Contents

Illustrations

Introduction

All musculoskeletal pain may be considered a sequela of soft tissue injury, irritation, or inflammation. Trauma in the broadest concept of the term is the greatest cause of soft tissue pain and functional impairment. Every aspect of the musculoskeletal system is subject to trauma in various forms.

The patient with musculoskeletal pain wanders from physician to physician in quest of relief or, at least, explanation and reassurance regarding his pain and dysfunction. Modalities are employed that have no rationale. Medications are prescribed to decrease pain and spasm while the cause remains unrecognized and, therefore, untreated. Unorthodox practitioners are sought and frequently, much to the dismay of the physician, afford the patient some relief. Nomenclature evolves and becomes ingrained in our culture, often to the disadvantage of the suffering patient. Insurance compensation and litigation compound the abuse of the patient.

In order to better understand the phenomenon of soft tissue pain and disability, one must find factors that relate pain and functional impairment in all the joints of the body. These factors develop along this line:

1. Functional anatomy of the involved segment of the body
2. Neuromuscular pattern of the moving parts
3. Tissue sites capable of eliciting pain
4. Responsible faulty neuromuscular mechanism

Each joint in the human body has its own characteristic structure and function and involves soft tissues. These soft tissues include muscles, capsules, ligaments, tendons, menisci, disks, and cartilaginous surfaces. All

1

these soft tissues are subservient to nervous system motor control, innervation for sensation and proprioception, and adequate blood supply. Functional anatomy clarifies the bioengineering principles of the moving parts in their normal movement or posture. The intricate movements controlled by the neuromuscular system are precise and allow little deviation. They are controlled by a coordinated central nervous system pattern that is partly developmental; they are modified by training and repetition.

Normal use and normal movement place no stress upon soft tissues. Excessive use, abuse, and misuse can cause irritation with resultant pain and disability. Posture or normal muscular activity may be altered by extraneous factors such as fatigue, distraction, anxiety, impatience, anger, and depression. External stress such as injury, trauma, or mechanical forces irritates the tissues that offer resistance. Once the neuromuscular pattern is altered, the normal biomechanics of the part is altered, and the soft tissues are abused and damaged. When these tissues contain pain-mediating nerves, pain results. Appreciation of the tissue sites so supplied with pain nerve endings affords meaningful evaluation of the pain mechanism as well as determination of the specific tissue site.

Therefore, it is mandatory that the functional anatomy of the involved structure be known in any musculoskeletal pain. Examination must specifically test the motion of that joint and expectantly reproduce specific symptoms. The history of the expected causative episode must be elicited also. The violation of the neuromuscular pattern that altered the biomechanics of the moving part must be clarified.

Once the abnormal pattern is apparent, treatment must restore the normal muscular pattern by training and repetition until the proper motion becomes automatic. Full range of motion of the impaired part must be restored. Whenever necessary or possible, the damaged tissues must be permitted to recover or be repaired. Many, if not most, musculoskeletal painful disabling abnormalities respond to treatment by mechanical means, and surgical intervention is usually not required nor effective.

CHAPTER 1
Soft Tissue Concept

In neuromusculoskeletal pain syndromes, soft tissue is the most frequent site of origin of nociceptive pain stimuli. Soft tissue is also the most common site of functional impairment of the musculoskeletal system. To understand fully the cause of pain and dysfunction, the term **soft tissue** must be understood, and its role in the causation of pain and disability must be clarified.

Soft tissues are the matrix of the body and are composed of cellular elements within a ground substance. The form(s) of the cellular elements are specified according to their intended function.

The four fundamental tissues of the body are

1. Epithelial tissue (skin) for protection, secretion, and absorption
2. Muscular tissue for contraction and motion
3. Nervous tissue for irritability and conductivity
4. Connective tissue for support, nutrition, and defense

All four of these basic tissues have unique structural and functional specializations, yet are intimately related and interdependent. With damage or alteration of any of these tissue elements, impairment and pain can result. It is such damage or alterations, either structural or chemical, that impairs function of the whole.

CONNECTIVE TISSUE

In the overall concept of soft tissue, the major component is connective tissue. Connective tissue includes muscles, tendons, ligaments, adipose tissue, bone, cartilage, blood, and lymph. Connective tissue is composed of various types of cells and fibers within a ground substance. The function of a specific connective tissue depends on the proportion and alignment of the cellular elements within the ground substance. These cellular elements are essentially collagen fibers (white), elastin fibers (yellow), and reticular fibers (or reticulum). The formation of connective tissue is illustrated in Figure 1–1.

Connective tissue is a highly complex tissue that supports, nourishes, and defends the organs of the body. It contains blood vessels, lymphatics, and nerves for its functions of defense, nutrition, and repair.

Classification and Function

There are numerous classifications of connective tissue, but the most simplistic is differentiation into **collagen**, which provides tensile strength, **elastin**, which provides elasticity, and **reticulum**, which provides support.

Collagen was defined by Virchow in the nineteenth century as "body excelsior" or "inert stuffing." Collagen consists of fibrils of formed elements contained within a ground substance that contains mucopolysaccharides. As most connective tissue contains varying percentages of and varying arrangements of collagen fibers, a collagen "fiber" merits descrip-

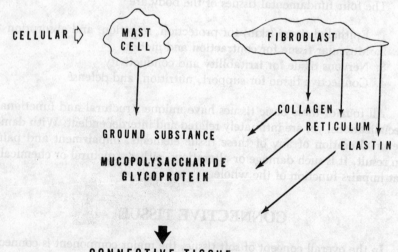

Figure 1–1. Formation of connective tissue. The mast cell secretes the ground substance, which contains mucopolysaccharide. This matrix ultimately contains collagen, reticulum, and elastin, which are derived from fibroblasts.

tion and definition (Fig. 1-2). A collagen fiber is a trihelix chain of approximately nine amino acids. These amino acids are connected to each other chemically and electrically to form a chain. At "rest" they are coiled, and can elongate by uncoiling. The extent of uncoiling determines the plasticity (rather than elasticity) of that collagen fiber. When the external uncoiling force is released, the collagen fiber resumes its coiled, resting length. The alignment and integrity of the collagen fibers determine the competence of the tissue to perform its precise function.

Connective tissue may be grouped as follows:

I. Connective tissue proper
   A. Loose connective tissue: contains many spaces in which are contained
      1. Collagen fibers, whose arrangement determines the tensile strength of the tissue
      2. Elastin fibers, the percentage of which determines the elasticity of the tissue
      3. Reticular fibers, which act in a supporting manner
   B. Dense connective tissue
   C. Regular connective tissue
      1. Tendon
      2. Fibrous membrane
      3. Lamellated connective tissue
II. Special connective tissue
   A. Mucous
   B. Elastic
   C. Reticular
   D. Adipose
   E. Pigmented
III. Cartilage
IV. Bone
V. Blood and lymph

Connective tissue has the following functional capacities:

1. Provides a supporting matrix for specialized organs.
2. Provides pathways for nerves, blood vessels, and lymphatics.
3. Facilitates movement between adjacent structures.
4. Forms bursal sacs to minimize friction and pressure.
5. Creates retraining mechanisms of moving parts by formation of bands, pulleys, and check ligaments.
6. Promotes the circulation of arterioles, capillaries, veins, and lymphatics.
7. Furnishes sites of attachment of muscles.

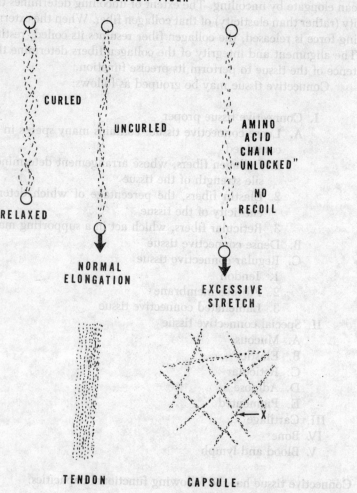

Figure 1-2. Collagen fiber. Each collagen fiber is a trihelix chain of amino acids bound together chemically (electrically). They uncurl to their physiologic length, then recoil when the elongation force is released. If the collagen fiber is elongated past its physiologic length, the amino acid chains become disrupted, and the fiber no longer returns to its resting length.

A tendon consists of parallel bands of collagen fibers. In a capsule, the collagen fibers crisscross and glide over each other at their intersection (X). The capsule depicted here elongates as far as each collagen fiber permits.

8. Aids in repair of injured tissue and repair by scar formation via its fibroblastic activity.
9. Forms fat storage space for conservation of heat and metabolites.
10. Contains histiocytes that furnish bacterial defense and immune reactions.
11. Participates in tissue nutrition via its tissue fluids.

Thus, connective tissue is vital to structural integrity and normal function. The arrangement of the cellular tissues determines function; disruption of structural alignment impairs function. Pain is a symptom of this disruption.

Types of Connective Tissue

Fascia. A fascia is essentially a sheath of connective tissue that envelops an organ system. The sheath surrounds the organ and adds tensile strength to that organ. A good example is a muscle bundle that is enclosed within a fascial sheath. The vertebral spine is enclosed within and supported by a fascial sheath. There is a layer of fluid between fascial sheaths that permits movement and minimizes friction by its lubricating action. Fascial sheaths are structurally highly specialized in their intended function (Fig. 1–3A).

Tendons. Tendons are bundles of parallel collagen fibers that principally connect muscle groups to the bones upon which they are to act (see Fig. 1–2). Tendons attach to bones via their periosteal sheaths, which are a type of connective tissue having a precise function.

Tendons have a glistening appearance, as they are invested in a loosely textured connective tissue envelope that contains a lubricating fluid similar to synovial fluid. The tensile strength of a tendon is tested when it is elongated. If the elongation force is excessive, it results in tearing the collagen fibers (disrupting the component amino acid bands) or avulsing the fiber attachment from the bony periosteum. Because a tendon is formed by parallel alignment of collagen fibers, it has only the elasticity of the individual component collagen fibers.

Ligaments. Ligaments are similar in structure and function to tendons except for the alignment of their collagen fibers. The collagen fibers of a ligament are more irregularly arranged than those of a tendon and contain, in addition to collagen, more elastin fibers.

Joint Capsules. A joint capsule, requiring specific flexibility, has loosely related collagen fibers that crisscross each other. This arrangement provides flexibility with fewer collagen fibers than are required in other tissues, such as tendons, in which the fibers are parallel (see Fig. 1–2). The fibers must also be free to glide upon each other at their intersections.

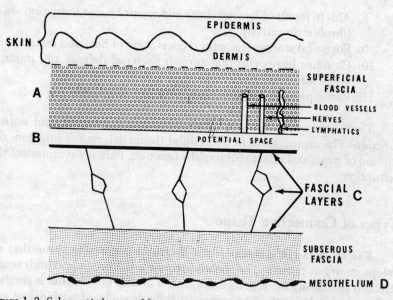

Figure 1-3. Schematic layers of fascia. *A,* Superficial layer containing fat, pressure nerve receptors, and blood vessels. *B,* "Potential" layer may become large space when dissected by extravasation or edema. *C,* Deep investing layers. *D,* Layers of pleura, peritoneum, pericardium, and so forth.

To preserve flexibility, capsular connective tissue also contains elastin elements and some fibrous elements. The percentage of each is vital in determining the exact flexibility of that specific capsule.

Muscles. There are two types of muscle fibers: involuntary smooth and voluntary striated. Muscles are considered to be specialized connective tissues with precise functions. Muscle fibers, held in bundles, are contained by a fascial type of envelope and terminate in a tendon to attach to bone periosteum.

Bone and Cartilage. Bone and cartilage are considered to be modified forms of collagen that contain large amounts of calcium. Cartilage is a form of connective tissue formed into a compressible hydrodynamic tissue that coats the ends of all bones related to joint structures. It contains abundant matrix supported by specialized collagen fibers. The matrix and the collagen fibers are formed by chondrocytes that secrete the mucopolysaccharide of the matrix and the collagen fibers.

BODY MECHANICS

Body mechanics is a simplistic term implying analysis of the moving parts of the body, that is, the neuromusculoskeletal system. Normal function and, therefore, normal body mechanics depend upon the integrity of the component soft tissues of the neuromusculoskeletal system.

Kinesiology is the scientific study of human movement. Kinesiology requires knowledge of structural skeletal anatomy, functional anatomy, and neurologic pathways.

To be appropriate, efficient, and pain-free, all musculoskeletal functions of the body must have a correct neuromuscular pattern encoded in the central nervous system. Initiation of a desired movement begins in the cerebral cortex of the brain and evokes a motor pattern in the central nervous system that is both voluntary and reflex (automatic). The muscular reaction, therefore, involves both the peripheral and central nervous systems, and both must be intact and properly oriented.

The muscles that implement motion originate and insert upon bones that form joints. The joints must be structurally normal, properly aligned, adequately flexible, and well lubricated. The muscles that activate these joints to achieve movement demand appropriate neurologic pattern functions.

Coordinated neuromuscular function has a genetic basis that becomes modified and refined by learning. The central nervous system is influenced by numerous stimuli that constantly bombard the motor aspect of the nervous system's components.

Neuromuscular function is a willed action with a desired, planned, and expected result. The cortical (brain) central nervous system pathways are voluntarily initiated and chemically and electrically mediated. The end result, the action, is a feedback mechanism, with the act being a learned activity.

Such learning, as a general concept, is the acquisition of response to these stimuli to control them, modify them, or extinguish them.

An animal brain and central nervous system at birth can function in a reflex manner, but growth, development, and learning modify these reflexes so that they gain functional precision and rapidity.

Learned neuromuscular activities can be considered skills. The learning of skills proceeds in three phases:

Phase One—Planning: Planning creates the objective, that is, the intended action, the goal of a desired function.

Phase Two—Execution: By performing the action with appropriate feedback, the action becomes more efficient and automatic. Less energy is required to perform the action, and less stress is imposed on the musculoskeletal system. The feedback of an accomplished action can be visual, tactile, proprioceptive, auditory, or even olfactory (Fig. 1–4).

Phase Three—Practice: Repetition of the correct, desired activity with creation of feedback that ascertains that the intended function

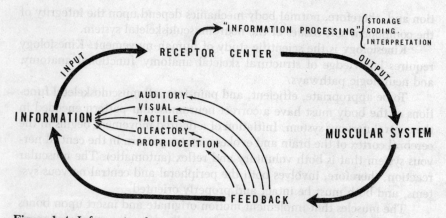

Figure 1–4. Information from internal and external environment via receptors is involved in the process of learning a skill. Almost all behavior is motor in nature; man responds with voluntary and involuntary movements, which include posture. The learning of skills proceeds in phases (See text). Feedback is one of the most important concepts in learning and is an important factor in the control of movement and behavior.

has been performed creates a correct habit that becomes encoded in the central nervous system. The musculoskeletal activities can now be initiated easily, efficiently, and automatically.

We learn by doing. We realize that we have done it "properly" by the feedback. We develop "skill" as we become more efficient and can perform with greater ease and less effort. Feedback is the mechanism that informs us that the desired, intended action has been developed, is effective, is pain-free, and requires minimal effort and concentration.

Neuromuscular function initially requires a conscious effort. Ultimately, implementation of the desired goal can be initiated automatically, using a learned, reflex, neuromusculoskeletal pattern. The feedback of this action is from the sensory end-organs involved in the action, for example, muscles, tendons, joint capsules, ligaments, and skin.

Initiating a specific motion and then developing a skill in the performance of the motion require practice and performance **without interruption** of any aspect of the neuromuscular patterns and feedbacks. Faulty neuromuscular activity occurring after a specific learned, skilled neuromuscular pattern has been encoded in the central nervous system can result from:

Poor physical conditioning: This primarily concerns poor muscular strength and endurance, articular capsule flexibility, and ligament elasticity.

Improper training: Learning faulty patterns and thus developing
faulty habits.
Fatigue
Emotional distraction
    Anxiety
    Anger
    Impatience
    Depression
External trauma

These interruptions can alter the normal, established neuromuscular
patterns and result in musculoskeletal injury. Functional impairment
results. Structural damage can result, further impairing function. From
this functional impairment can result initiation of nociceptive stimuli with
resultant **pain.**

Every musculoskeletal system is dependent upon intact neurologic
patterns, proper skeletal anatomic alignment, well-conditioned soft tissue
of the skeletal system, and minimization of the distractions listed previously.

Joint Motion

Joint function, so vital in proper neuromuscular motion, is a coopera-
tive venture between engineering and medicine. The type and extent of
normal movement occurring in any joint depends upon (1) the form of the
articulating surfaces, (2) the restraining influence of the articular capsules
and ligaments, (3) the atmospheric pressure within the joint, and (4) the
control exerted by the muscles that interact with that specific joint.

A typical joint is formed by two opposing bones whose articular surfaces are coated with cartilage, enclosed within a capsule containing synovial fluid, and moved by muscles attached via tendons.

Movement of the opposing articular surfaces depends on the shape of
those surfaces. Movement of a joint can thus be termed **displacement.**
There are basic types of displacement (joint movement): (1) true spin
about an axis, (2) arc slide, and (3) spin about an axis with rotation about
a fixed point of a surface (Fig. 1–5).

There are basic types of joints, according to the shapes of the opposing
surfaces, that dictate their specific functions:

1. **Plane joints,** in which the opposing surfaces are flat. These are
   essentially weight bearing with limited motion, and what motion
   occurs is that of slide or shear.
2. **Spheroid joints,** in which the opposing surfaces are rounded, one
   concave and one convex. They are articulate about a central axis
   "about each other" as in the so-called ball-and-socket joint.

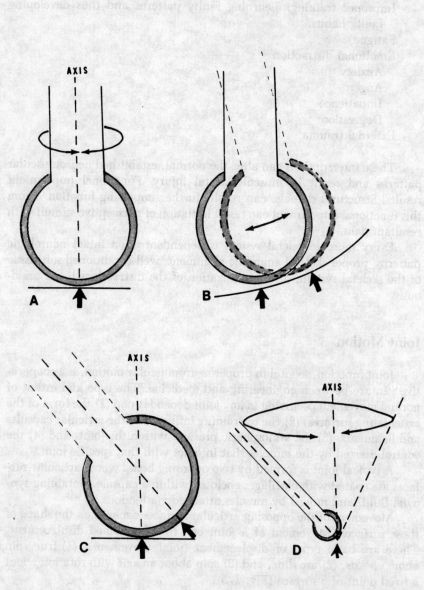

Figures 1-5. Basic kinds of displacement: *A*, True spin about an axis. *B*, Arc slide called "swing" when there is no simultaneous rotation or spin. *C*, Spin about the axis and rotation. *D*, Spin is rotation about an axis perpendicular to fixed surface. (Adapted from Licht, S (ed): Arthritis and Physical Medicine. Williams & Wilkins, Baltimore, 1969.)

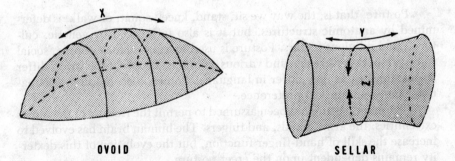

Figure 1-6. Joint surfaces are considered to be of two basic types: ovoid and sellar. Ovoid permits motion in one plane, X; sellar permits motion in two planes, Y and Z.

3. **Cotyloid joints,** also spheroid joints, in which the articulating surfaces are relatively longer in one direction.
4. **Hinge joints,** in which movement is usually limited in one plane. There may be some slight rotation within these joints, but the prime movement is in one direction.
5. **Condylar joints,** in which the two articulating surfaces permit a hinge-type movement.
6. **Trochoid joints,** with rotatory movement of a ring-shaped structure about a bony pivot (for example, atlas about the axis in the cervical spine).
7. **Sellar joints,** in which one of the articulating surfaces resembles a horse saddle.

To simplify joint movement, these have been reduced to two basic types of surfaces: ovoid and sellar (Fig. 1-6).

Throughout the text, joints of a specific type, structure, and function will be discussed as each portion of the human body is considered.

Posture

Posture affects every part of the musculoskeletal system of the body and thus merits separate consideration.

The posture of mankind is unique among animals. Erect stance and ambulation of man are precarious. Painful disabling conditions of the soft tissues of the musculoskeletal system are directly or indirectly related to posture in standing, walking, moving, lying, sitting, bending, or lifting. Posture is directly and indirectly influenced by genetics, training, bodily conditioning, and the emotions. All are significantly interrelated.

Posture, that is, the way we sit, stand, kneel, squat, or walk, is determined by anatomic structures, but it is also influenced by genetic, cul-
tural, and religious factors. Posture is used to express, for example, social
status, respect, reverence, and various emotions. People of the world differ
in posture, just as they differ in language, clothes, diet, housing, occupation, and even musical preference.

The erect posture has been assumed to permit the free use of the upper
extremities: the arms, hands, and fingers. The human brain has evolved to
increase the skill of hand-finger function, but the evolution of this dexter-
ity remains dependent upon the erect posture.

Standard posture has been defined by the Posture Committee of the
American Academy of Orthopaedic Surgery (1947) as "skeletal alignment
refined as a relative arrangement of parts of the body in a state of balance
that protects the supporting structures of the body against injury or progressive deformity." The "skeletal alignment" in this definition refers to
more than merely the components of the vertebral column. It refers to all
the musculoskeletal components of the body that are connected to the vertebral column.

Ordinary posture is normally considered as standing erect with the
arms hanging loosely at the sides. Sitting, as in a chair, is not a posture that
is practiced world-wide. One quarter of the human race habitually
"rests," (i.e., relieves the body of the effects of gravity), by squatting. People of different cultures "squat" in different ways.

The human body is poorly engineered for standing upright. Geome-
trically, the body is shaped with heavy wide parts balanced on narrow
bases (Fig. 1–7). Soft tissue components must support this awkward structure.

In addition to this structural aspect, the erect body has four major
curves that must be balanced upon a central axis of gravity. Because of its
physiologic composition and its engineering, the body must be supported
with minimal stress and minimal energy expenditure. The head must be
supported at the center of gravity based upon all the vertebral structures to
the feet. All the curves of the vertebral column and the triangles of the
extravertebral structures must be centered as close to the center of gravity
as possible (Fig. 1–8).

A person's posture in standing, sitting, walking, bending, lifting,
pushing, or pulling influences all the functions of all the extremities. Posture is influenced by structural alignment and the structural integrity of
the component parts. The soft tissues that add their support to this alignment must have adequate strength, endurance, and flexibility, and must
be well coordinated neuromuscularly. Training must ensure this proper
coordination. Conditioning must ensure the efficacy of the soft tissues and
the elimination or minimization of all interfering factors that adversely
influence proper function.

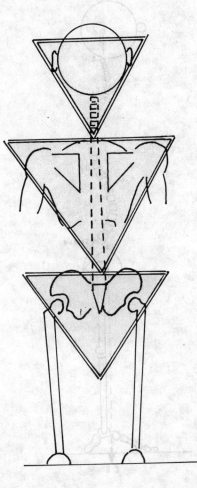

Figure 1-7. Body engineering of stance performed with broad heavier parts at the top situated upon a narrow base.

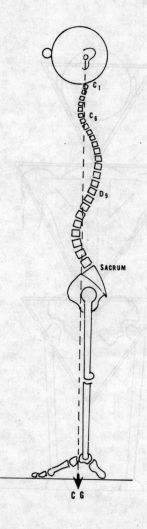

Figure 1–8. Erect posture through the center of gravity. It passes through the external meatus, through the odontoid process, slightly posterior to the center of the hip joint, slightly anterior to the center of the knee, and slightly anterior to the lateral malleoli.

BIBLIOGRAPHY

Becker, RF: The meaning of fascia and fascial continuity. Osteopathic Annals 3(3):38–42, Feb 1975.

Cailliet, R: Mechanism of joints. In Licht, S (ed): Arthritis and Physical Medicine, vol 11. Elizabeth Licht, 1969, pp 17–34.

Goltz, RW, Rodgers, EG, and Ashe, BM: Connective tissue disorders: A mystery from nascence to senescence. Modern Medicine, March 4, 1974, pp 31–37.

MacConaill, MA: The movements of bones and joints. J Bone Joint Surg 32-B(2):244–252, May 1950.

Otaka, Y, and Watanabe, Y: Pathology of connective tissue disease. In Otaka, Y (ed): Biochemistry and Pathology of Connective Tissue. Igaku Shoin Ltd., Tokyo, 1974, pp 152–179.

Prockop, DJ, Kivirikko, KI, Tuderman, LT, and Guzman, NA: The biosynthesis of collagen and its disorders. New Engl J Med 301(1):13–85, July 1979.

Radin, EL, and Paul, IL: A consolidated concept of joint lubrication. J Bone Joint Surg 54-A(3):607–616, April 1972.

Stravino, VD: The synovial system. Am J Phys Med 51(6):312–319, 1972.

Thompson, RC, and Robinson, HJ: Current concepts review: Articular cartilage matrix metabolism. J Bone Joint Surg 63-A(2):327–331, Feb 1981.

CHAPTER 2

Pain From Soft Tissue: Its Mechanisms and Evaluation

Pain can no longer be considered merely a symptom. It may actually be a disease. How an evaluator thinks about pain depends on his or her learning, expertise, or specialty. To a neurologist or a neurosurgeon, pain is the result of neurologic abnormality. To an orthopedist, pain is the result of musculoskeletal deviation. To a psychiatrist or psychologist, pain is an emotional reaction to a physical insult. The behaviorist considers pain a process of manipulating the environment or the injured person's relationships. The organically oriented physician may consider pain as organ language that indicates an organic pathologic condition needing "specific" treatment or surgical eradication.

Steinbach[1] classified pain as an abstract concept in which the patient describes a personal sensation of "hurt", which may signal tissue damage, and which may serve to protect the person from further harm.

Regardless of the interpretation of the symptom of pain—be it neurophysiologic, physiologic, psychologic, or behavioral—noxious irritation of tissue causing nociceptive stimuli plays a major role in initiating the ultimate implementation of the symptom of pain.

The symptom of pain is considered to originate in soft tissues, which implies that there are neural pathways that are initiated from these soft tissues. The subject of pain is broad and has been receiving a great deal of interest and study in recent decades. It can be safely stated that *pain* is considered to be an unpleasant sensation of an acute or chronic nature— that is, differentiated by the factor of time.

Acute pain is termed nociceptive, with the implication that an abnormal stimulation of a sensory nerve in the peripheral tissues originally initiates a sensation that will ultimately be interpreted as pain. The definition of chronic pain varies, depending on the school of thought of the person describing the condition. Chronic pain is currently considered to be an abnormal, unpleasant, and possibly disabling sensation that has persisted for more than 6 months, has defied response to meaningful treatment, and usually, has involved symptoms persisting "after" the nociceptive stimulus is operant.

The subject of chronic pain fills volumes of literature and has been greatly researched. It is not the purpose of this book to discuss chronic pain in its entirety, but insofar as pain often originates from the soft tissue of the neuromusculoskeletal system as nociceptive stimuli, the original site and mechanism of the production of pain merit discussion.

PAIN: ITS ANATOMIC BASIS

Pain is a subjective response to injury of, or insult to, a tissue of the body. A nerve must be involved in the transmission of this impulse from the injured tissue to the spinal mechanisms that ultimately are recorded and interpreted as pain by the brain. These nerve pathways originate in organ systems, of which the musculoskeletal system is predominant.

Initiation of nerve conduction is electrochemical. It is considered to be molecular, whereby there is passage of sodium, potassium, and chloride through a semipermeable membrane that surrounds the axon of a nerve. This passage of ions changes the electrical status of the axon, and this alteration depolarizes the nerve, resulting in the transmission of an impulse.

This description of the transmission of a nerve impulse is simplistic. It does not explain the transmission of the sensation of pain. Further evaluation of the interpretation of a nerve impulse as pain is necessary.

Pain is a subjective experience based on an interplay of biologic, psychologic, and social factors. The initiation of a nerve impulse that is ultimately interpreted as pain resides in the soft tissues in many instances, and the acute pain impulse can be eliminated, modified, and corrected if its mechanism is understood. It is probable that interception, modification, and even prevention of the pain impulse from soft tissue in its initial nociceptive transmission may prevent prolonged disability and even abort progression into chronic pain.

Information initiated by a noxious stimulus is transmitted through peripheral nerves, which then synapse to nerve pathways of the spinal cord, and the information ascends by way of numerous pathways to the brain, continuing via the medulla and midbrain and terminating in the thalamus. From the thalamus, the information travels to the cortex, where its "meaning" is interpreted.

Nociceptive stimuli originate in the soft tissues for transmission via peripheral nerves. The old concept that specific end-organs carried specific sensations has been generally refuted, but there remain certain generalizations that deserve consideration.

NOCICEPTIVE RECEPTIVE MECHANISMS

Stimuli such as heat, cold, touch, and pressure were considered to be "picked up" by specific end-organs. These end-organs were later labeled as Krause's bulbs, Ruffini's heat-sensor organs, Meissner's touch corpuscles, and so forth. These end-organs allegedly transmitted the "specific" sensations (e.g., heat or touch) and were all connected to peripheral nerves that subsequently transmitted these sensations.

Research has refuted this specificity of end-organs and nerve fiber transmission. Another theory replaced this concept. This was the theory of "intensity," in which the sensation of pain was thought to result from an intensity of stimulus beyond a certain threshold. It was thought that certain intensity of stimulus was perceived as "touch," and that increasing or prolonging the intensity beyond that threshold initiated an unpleasant sensation consistent with pain. A theory involving both elements—specific end-organs and nerves, and appropriate stimulus intensity and duration of that intensity—is currently accepted.

There are three major classes of pain receptors specifically related to the previously mentioned stimuli (Fig. 2–1):

1. **Mechanoreceptors** have a high threshold and respond to nondamaging mechanical stimuli. Their impulses are transmitted via A delta fibers in the peripheral nerves.
2. **Thermoreceptors** respond to changes in temperature and initiate impulses in the A delta fibers as well as in the unmyelinated C fibers.
3. **Polymodal receptors** respond to potentially damaging stimuli such as chemical, high thermal, or strong mechanical stimuli. They create impulses transmitted via C fibers.

Tissue injury, regardless of its type, liberates a number of chemical substances, which include histamine, serotonin, bradykinin, potassium ions, and prostaglandins. These chemicals stimulate smooth muscles in the vicinity of the traumatized tissue via sympathetic fiber endings. The stimulated sympathetic fibers further produce vasomotor changes, which include ischemia and edema.

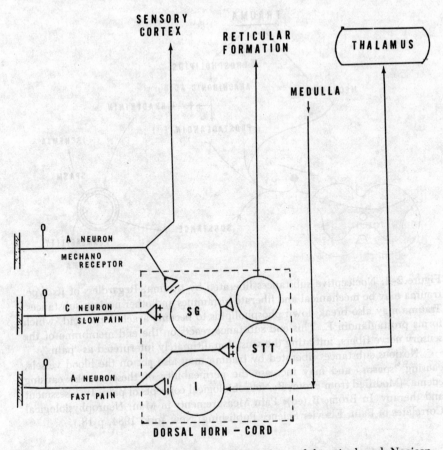

Figure 2-1. Afferent (sensory) fibers in the dorsal horn of the spinal cord. Nociceptive stimuli are transmitted via three types of peripheral nerve axons: A beta (mechanoreceptors), C neurons (transmitting slow pain sensations), and A gamma (transmitting fast pain sensations). They synapse with fibers of the substantia gelatinosa (SG) and enkephalinergic interneuron (I) and ultimately ascend via the spinothalamic tract (STT) to the thalamus. There are numerous interconnections that ascend to the cortex via other routes and via the reticular system. (Modified from Bond, M: Pain: Its Nature, Analysis and Treatment, ed 2. Churchill Livingston, Edinburgh, 1984, p 26.)

Among these chemical substances is the fatty acid arachidonic acid, which is liberated from phospholipids by trauma to form prostaglandins (Fig. 2-2). The major forms of prostaglandins currently considered to be major nociceptive agents are PGE_2 and $PGF_{2\alpha}$. It is interesting that it is this chemical stage of prostaglandin formation that is interrupted by steroids and salicylates in the relief of pain.

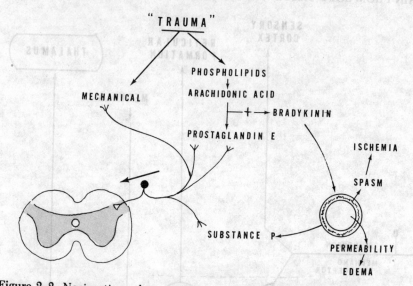

Figure 2-2. Nociceptive substances liberated by trauma. Regardless of its type, trauma may be mechanical and liberate histamines and other noxious substances. Trauma may also break down phospholipids to become arachidonic acid, which forms prostaglandin E. This end substance reacts on the end membrane of the sensory nerve fibers, initiating the sensation ultimately interpreted as "pain."

Noxious substances liberated by trauma may also act on the blood vessels, causing "spasm" and may increase the permeability of these vessels, causing edema. (Modified from Brom, B: Neurobiological concepts of pain: Its assessment and therapy. In Brom, B (ed): Pain Measurements in Man: Neurophysiological Correlates of Pain. Elsevier Science Publishing, New York, 1984, p 18.)

The old adage of tissue injury causing dolor, calor, tumor, and rubor, postulated centuries ago by Celsus and later by Virchow, can now be explained chemically.

Dolor (pain) is initiated by the liberated kinins, histamine, potassium, and prostaglandins acting upon the nerve endings. **Calor** (heat) can be explained by the liberation of histamine-like substances at the tissue site from sympathetic nerve stimuli, causing vasodilatation. **Tumor** can be explained by the formation of edema and inflammation, which cause chemical irritation as well as mechanical noxious stimulation. **Rubor** is redness from excessive blood flow to the part.

The fifth phase of tissue inflammation postulated by Galen after dolor, rubor, and calor was "functio laesa," which means "loss of function." This loss of function from tissue injury explains mechanical impairment as well as disability from pain.

Noxious chemical substances affect the sensory nerve endings that eventually transmit impulses to the central nervous system, ultimately resulting in pain. These are the same sensory nerves that transmit the sensations of heat, cold, pressure, and touch. At certain intensities, the chemical stimuli are non-noxious and are interpreted as their specific sensations.

Regarding the sensation of pain, it can thus be postulated that the sensory receptors are involved in *any* sensation, but that it is the **intensity**, the **type**, and the **duration** of the stimulus that may initiate pain, as was postulated by Wall in his concept of pain.

Afferent A delta and C nerve fibers are known to carry sensation in the sensory nerves to the dorsal ganglion. It has been postulated that 30 percent of the "motor" nerves contain sensory fibers that carry sensation to the spinal column gray matter.

There are also sensory fibers in the sympathetic (autonomic) nervous system. Both somatic and sympathetic sensory nerves interact within the posterior root ganglia. This interaction possibly constitutes another "pathway" in the hypothalamic limbic pain circuit in the dorsal column. It is also a pathway in the transmission of "visceral pain" from internal organs (Fig. 2-3).

Although both systems ascend similar spinal cord routes, they deviate as to their final destiny and the sensations they carry. At their origin at the nerve receptors within the soft tissues, the sympathetic nerves are possibly stimulated by norepinephrine and acetylcholine, both of which are released by the damaged tissues along with kinins, histamines, potassium ions, and prostaglandins. These sympathetic stimuli act locally and mediate impulses to cause local vasodilatation.

There are fibers in the peripheral nerves that specifically transmit the sensation of pain. There are fibers that predominantly transmit other sensations, such as touch, temperature, position sense, and so forth, but if the stimulus is adequately strong or prolonged, any of these fibers can carry a sensation that is ultimately unpleasant and is interpreted as pain.

Proponents of a newer concept of pain transmission and interpretation, Melzack and Wall[2] proposed a neurophysiologic basis for pain with the nerve impulses proceeding from the periphery and being modulated at the spinal cord level. The "gate theory," as this modulation theory was termed by Melzack, conceded the following:

1. There is a functional specificity in the central nervous system.
2. There is a temporal factor as well as a summation of the impulses that ultimately generate pain.
3. There is included in the pattern the intervention of the emotions, previous experiences, and behavior.

The gate theory assumed that a noxious stimulus initiates an impulse in the peripheral nerve, which ascends to the spinal cord, where it is "modulated" in the dorsal columns. The region in the dorsal column of this

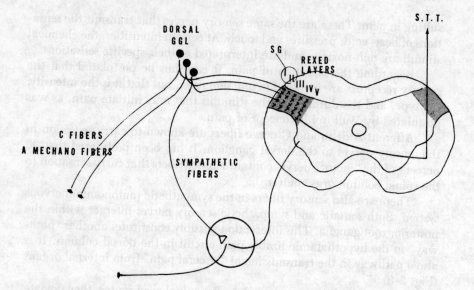

Figure 2–3. Causalgic (autonomic) transmission of pain sensation. Trauma irritates somatic afferent C fibers, A mechanofibers, and sympathetic fibers, whose impulses proceed to the dorsal column of the cord. There is a cord interneuron connection that transmits efferent impulses via the autonomic system to the periphery, which sensitizes the skin to mechanical (light touch) input.

At the cord, the afferent fibers initiate neuronal activity in the Rexed layers of the dorsal column. The Rexed layers I and II are the substantia gelatinosa. (Modified from Roberts, WS: A hypothesis on the physiological basis for causalgia and related pains. Pain 24:297, 1986.)

modulation was designated to be in the area of the substantia gelatinosa (SG). More recently, it has been further specified by Rexed to be in the region he designates as I–II. The area I–II is therefore the area of impulse modulation.

In the SG, there are T cells that transmit and allegedly modulate the impulses of the noxious stimuli (Fig. 2–4). The T cells are activated by these noxious impulses. The T cells are further modulated by other impulses emanating from the SG from other peripheral impulses and from impulses descending from the higher levels such as the median raphe, hypothalamus, and limbic system (Fig. 2–5).

The peripheral nerve fibers that "carry" impulses capable of causing pain vary in speed of conduction. The C fibers are of small diameter and are very thinly myelinated. They conduct impulses slowly and are considered to be "transmitters of slow pain." Slightly larger fibers with a myelin sheath carry impulses faster and are termed A delta fibers. These latter

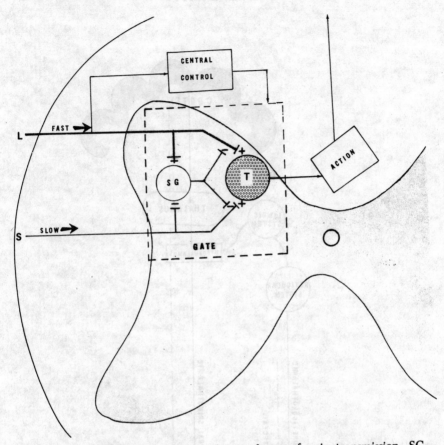

Figure 2-4. Wahl-Melzak concept of gate theory of pain transmission. SG—substantia gelatinosa; T—T cells.

fibers respond only to an intense stimulus and supposedly carry sensations of acute injury such as a pinprick to the skin.

Once the nerve impulses reach the dorsal column of the cord, they enter lamina I and V fibers and, to a lesser degree, some lamina II fibers. Lamina III, IV, and VI fibers probably do not carry noxious stimuli. Lamina I and V fibers carry both noxious stimuli and non-noxious stimuli, and lamina II fibers carry only noxious stimuli, thus transmitting the sensation ultimately perceived as "pain."

The basic concept of the gate theory is that pain sensation carried by slow fibers and initiated by a noxious stimulus activates the T cells and proceeds further to the thalamic system to be recorded and interpreted as pain. The SG can modulate this impulse transmission and "block" the T cell—that is, the gate can be closed.

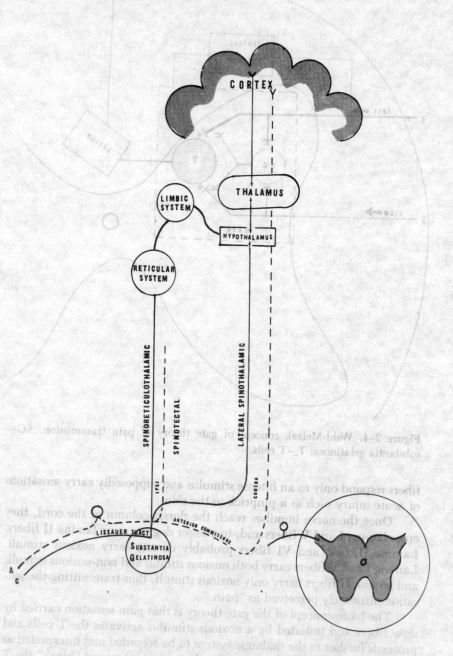

Figure 2–5. Two major neurophysiologic transmission systems. Spinothalamocortical system (A) has spatiotemporal localization; spinoreticulothalamic system (C) has no localization but is involved in emotional (limbic) and avoidance reaction.

In the absence of a noxious stimulus, there is a good balance of the transmission of impulses. When T cells are bombarded by high-intensity stimuli or large volumes of noxious stimuli, pain can be transmitted to the spinal cord, thalamus, and cortical system.

Stimulation of the SG can alter the sensitivity of the T cells and "modulate" the impulse transmitted. There will be further discussion of the modification of pain and the establishment of a rationale for the various methods used in the treatment of pain.

The noxious stimulus that initiates the impulse in the peripheral nerve is probably the release of chemical substances that occurs at the end-organs of the nerves and lowers the membrane potential, causing the nerve to transmit an impulse. The noxious stimuli are liberated by trauma to soft tissue, and they activate the C neurons directly.

These noxious stimuli include substance P, bradykinins, prostaglandin precursors, and numerous others. In essence, the injured tissue "insults" the end-organ of the peripheral nerve and initiates the ascent of an impulse.

The impulses transmitted via the sensory nerves enter the dorsal horn of the gray matter, where they ultimately stimulate internuncial neurons in their ascent to the spinothalamic and spinotectal tracts of the cord cephalad. In the cord, the spinothalamic tract receives impulses predominantly from lamina I and V, and to a lesser degree lamina II, of the Rexed cells. From these cells, the impulses enter internuncial fibers that cross over the midline. The impulses then ascend the anterolateral fasciculus of the cord to the thalamus and midbrain via the spinothalamic and spinotectal pathways.

As the impulses approach the thalamus, the nerves send collateral fibers to the periaqueductal and periventricular gray matter of the midbrain. It is currently considered that the lateral thalamus receives sharp localized pain stimuli whereas the medial thalamus receives diffuse, poorly localized, "burning" sensations. There are other sensations besides pain that are transmitted via the spinothalamic tracts.

There are descending pathways that interplay upon the dorsal horn. The major ones originate in the periaqueductal gray (PAG) area of the midbrain. Cells within the PAG are rich in endogenous opiates. These opiates are released to activate the nucleus raphe magnus (NRM) in the midbrain, which in turn releases serotonin. This serotonin releases opiates throughout the central nervous system as well as upon the dorsal horn. The area within the dorsal horn is the "gate" specified previously.

As the NRM has been found to contain serotonin, substance P, and somatostatin, it is probable that the NRM substance both inhibits and excites the gate in the dorsal column, thus influencing the sensation of pain.

Three opiates have been isolated from the hypothalamic region (the NRM): (1) β-endorphin, (2) enkephalin, and (3) dynorphin.

1. β-Endorphin is located primarily in the pituitary gland and in the hypothalamus. The axons from the hypothalamus that traverse the PAG contain the endogenous beta endorphin, which is released from stimulation of the PAG in the production of analgesia.
2. Enkephalins are universally distributed throughout the central nervous system and act on local nerve circuits.
3. Dynorphin is mostly concentrated in the PAG and is considered a powerful analgesic.

All of these factors are located within the hypothalamus, the limbic system, the PAG, and so forth, and they are related to the dorsal gray matter; it becomes apparent that the *descending* fibers to the gate of the dorsal gray matter of the cord are significantly involved in the perception of pain. As the hypothalamus and limbic system are also intimately related to the emotions, the relationship between pain and the emotions becomes more clear (Fig. 2–6).

Psychologic factors have an impact on the nociceptive impulses originating from the peripheral tissues. Spinal sympathetic reflexes can affect the circulation of the tissues surrounding the nociceptive nerve fibers that become involved in local injury.

Epinephrine is released at these sympathetic nerve endings, and it sensitizes as well as activates the peripheral neuroreceptors. Sympathetic arousal of the higher brain centers, such as that resulting from emotional stress can exacerbate the nociceptive impulses originating from soft tissue injury.

Emotional stress has been shown to release adrenocorticotropic hormone (ACTH) and β-endorphin, both of which interact with enkephalins. The endocrine involvement of the enkephalin serotonin explains the benefit derived from stress management programs and from physical exercise.

The pathways from the periphery have been traced to the thalamus, the hypothalamus, the limbic system, and the midbrain. From there, they ascend to the cortex, where the sensory impulses are interpreted and evaluated as to their significance. These latter pathways are as yet unclear, but it is apparent that at the cortical level, past experiences and psychologic and social factors play a part in grading the extent and significance of the sensation of pain.

NOCICEPTIVE SITES OF SOFT TISSUE PAIN

Pain from soft tissues initiates reaction at the end-organs of sensory nerves. An understanding of the specific tissue site for these nociceptive impulses is necessary to understand the meaning of the history, the basis for

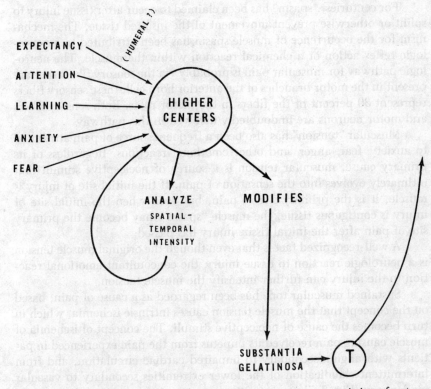

Figure 2–6. Higher-center modification influencing gate transmission of pain.

the examination, the rationale for the laboratory tests, and ultimately, the basis for the treatments rendered.

The many kinds of soft tissue have been enumerated in Chapter 1, and the neurologic pathways of pain transmission have been described in this chapter. The specific tissue merits consideration. Which tissue is the greatest offender depends on the type of injury and on the extremity or the anatomic site.

Muscles

Of the numerous soft tissues that can be the locus of pain, muscle is most frequently involved. Pain can originate primarily from muscle tissue, or secondarily from the contiguous tissue that has sustained injury, with secondary muscle "protective spasm." These contiguous tissues may be joints, bones, periosteum, ligaments, or tendons.

For centuries, "spasm" has been claimed to occur after tissue injury to splint or otherwise prevent movement of the insulted tissue. The mechanism for the occurrence of muscle spasm has been attributed to a neurologic reflex action or a chemical reaction within the muscle. The neurologic pathway for muscular pain is probably via the sensory fibers that are present in the motor branches of the anterior horn cell; these sensory fibers represent 30 percent of the fibers in the motor roots. Both sensory fibers and motor neurons are undoubtedly involved in this pathway.

Muscular "tension" has also been a frequent source of pain attributed to anxiety, fear, anger, and other emotional reactions. Regardless of its primary cause, muscular tension is a source of nociceptive stimuli that ultimately evolves into the sensation of pain. If the initial site of injury is muscle, it is the primary site of pain, but even when the initial site of injury is contiguous tissue, the muscle "spasm" may become the primary site of pain after the initial tissue injury has passed.

A well-recognized fact is that even though the original muscle tension is a neurologic reaction to tissue injury, the concomitant emotional reaction to the injury can further intensify the muscle tension.

Sustained muscular tone has been regarded as a cause of pain, based on the concept that the muscle tension causes intrinsic ischemia, which in turn becomes the cause of nociceptive stimuli. The concept of ischemia of muscle causing pain received its impetus from the pain experienced in patients with angina secondary to impaired cardiac circulation, and from intermittent claudication of the lower extremities secondary to vascular inadequacy (Fig. 2–7).

Variations of arterial pressure have been considered to influence the severity and rapidity of onset of muscular pain. An exercise performed with the arm held above the cardiac level, a position that elevates the arterial pressure in the arm, was found to cause earlier and more severe "ischemic" pain than a similar exercise done with the arm held at the horizontal level.

Studies were conducted in which exercises were performed with a tourniquet applied to the arm above systolic pressure as an example of exercise done under conditions of vascular occlusion. Repeated exercises of sustained muscular contraction performed by the arm without the tourniquet ultimately caused muscular pain. Significantly, the same exercise performed with a proximal tourniquet caused the same pain after fewer contractions.

The interpretation was that with the tourniquet, there were ischemia and an accumulation of metabolites from muscular contraction causing the pain.

This concept of ischemic pain was subsequently refuted when prolonged ischemia from a tourniquet *without* exercise did not cause pain,

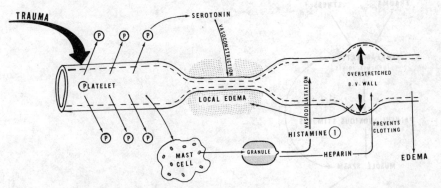

Figure 2-7. Schematic concept of the effect of trauma, a vasochemical reaction.

and the application of a tourniquet to the arm *before* sustained exercise did not cause more pain from subsequent sustained exercise.

After exercise with an applied tourniquet, removing the tourniquet did not permit the muscle to be re-exercised without pain. These experiments implied that it was probably the accumulation of metabolites, and not the ischemia, that caused pain, and that the metabolites remained within the muscle after exercise to initiate nociceptive pain even though the blood flow returned. Ischemia undoubtedly plays a part in muscular pain, but this is probably because the ischemic muscle contraction causes a change in the resultant metabolites. It has been an accepted fact in sports medicine that a strenuously exercised muscle group must be rested, not exercised, the next day; otherwise, pain and a lack of strength and endurance may occur.

The specific metabolites have not been identified. Originally, these were considered to be lactic acid or pyruvic acid, which are end-metabolites of the anaerobic aspect of muscular contraction. The belief that lactic acid is involved has been refuted, however, as patients with McArdle's disease—hereditary absence of muscle phosphorylase preventing the production of lactic acid—develop muscular pain from sustained exercise.

While there is not yet a scientific explanation for painful muscle spasm—be it prolonged exercise with excessive production of metabolites, prolonged accumulation of these metabolites, ischemia, or other factors—it is an accepted fact that muscular pain does occur (Fig. 2-8).

Isolated painful sites in muscle, in which there are palpable tender muscular "nodules," have been clinically identified as "trigger points," or as resulting from conditions variously labeled as myofascial pain syndromes, fascitis, myofascitis, fibrositis, muscular rheumatism, and psychogenic myalgia. These as yet unclarified muscular painful nodules are a

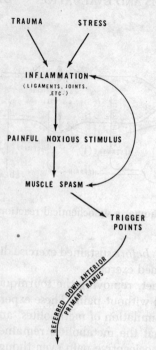

TRAUMA   STRESS

INFLAMMATION
(LIGAMENTS, JOINTS,
ETC.)

PAINFUL NOXIOUS STIMULUS

MUSCLE SPASM

TRIGGER
POINTS

REFERRED DOWN ANTERIOR
PRIMARY RAMUS

Figure 2–8. Schematic concept of pain manifestation. Trauma to joints, ligaments, and other soft tissues ultimately creates trigger points that refer pain to distal sites.

frequent clinical entity leading to pain, impairment, disability, and frequently, therapeutic frustration.

As early as 1843, Froriep[4] labeled the condition involving tender bands or cords within muscles as "rheumatism," and he considered these tender palpable bands to be deposits of connective tissue. Shortly thereafter, Grauhan[5] histologically described these bands as fibrous tissue that surrounds muscle fibers and that is infiltrated with lymphocytes. Steindler and Luck[6] termed these "trigger zones" and considered the sites to be within the ligaments or a muscle.

Travell[7] conducted many studies on trigger points since 1942 and reported that her patients revealed exquisite tender spots, which when pressed referred pain to a predictable distal site. She and her coworker Simon[8] wrote extensively on the subject of myofascial pain syndromes.

The literature specified criteria for making this diagnosis: (1) exquisite tenderness over (2) a circumscribed, painful, thickened mass and (3) pressure on that point causing pain referred to a distal muscle site. The history usually revealed that the cause of trauma was positional, mechanical, postural, or emotional. Travell[9] claimed benefit from spraying the trigger points with a vasocoolant followed by stretching of the involved muscle.

Kraft, Johnson, and LeBan[10] claimed that four criteria were needed to justify the diagnosis of myofascial pain syndrome: (1) a "jump" sign, whereby the patient allegedly jumped when the trigger point was palpated; (2) the "rope" muscle sign, whereby a rope-like swelling could be palpated by the examiner's finger passing over the involved muscle; (3) vasomotor signs, such as blanching followed by hyperemia after actively shaking the extremity; and (4) relief of symptoms by local infiltration of an anesthetic agent into the "trigger" zone.

For centuries, muscular overactivity has been documented as causing painful disabling musculoskeletal syndromes. Postural, occupational, emotional, and metabolic traumata have been implicated, yet many conditions are still not clearly understood. Modalities of physical medicine remain the method of choice for eradicating symptoms and reducing the disability.

Joints

The tissues of joints are also major sites of nociceptive stimuli. By definition, joints are anatomic regions where two bones meet to create a movable articulation. The opposing bones are usually coated by cartilage and are surrounded by a flexible capsule. A layer of synovial cells lines the capsular surfaces.

To minimize frictional wear and tear, joints are lubricated by an exudate from the cartilage and by synovial fluid secreted by the synovial cells.

Cartilage is normally avascular. It receives its nutrition by imbibition from alternate compression and relaxation (expansion). This compression-expansion causes secretion of mucin and hyaluronidase, which enter the joint capsule and coat the surface of the cartilage with a slippery substance. Upon expansion of the cartilage, as the compressive force is released, the joint fluid re-enters the cartilage as its nutrient. This action is mechanical and is enhanced by the matrix of the cartilage reinforced by collagen fibers. The blood supply that furnishes the synovial and cartilage fluid enters from periarticular sites.

Normal cartilage has neither blood supply nor nerve supply. There are no somatic or sympathetic nerves in the tissues of the joint space, except for the sensory nerve ending and vasomotor nerve to the capsule. Thus, the joint is sensitive, but the normal cartilage is insensitive.

The normal joint has a capsule that is connective tissue, which allows flexibility and contains the enclosed synovial fluid. Many joints have a thickening of the capsule, which forms a ligament in essence. The purposes of these capsules and enclosed ligaments are to limit range of motion and afford stability of the particular joint. Each specific joint of each extremity will be considered in subsequent chapters.

Joint damage creates pain and impairment (1) from direct trauma such as a mechanical external stress, (2) from intrinsic trauma such as faulty use or overuse, or (3) from anatomic malalignment. Infection, inflammation, and metabolic joint disease may be added as causes of pain and impairment.

Pain from a joint must come from the capsule or the ligamentous thickening of that capsule, which as a result of trauma exceeds the limits of flexibility. The nerves of that capsule become the mediators of nociceptive stimuli. Pain is not caused by damage to the cartilage per se, unless the capsule is involved, or unless there is denuding of the cartilage exposing the bone underneath, in which case the bone becomes the site of nociceptive stimuli.

Nerves

In discussions of pain, the term "nerve pain" always emerges. It is a truism that pain requires a nerve for its transmission; thus, pain from nerve damage is difficult to clarify.

Pressure on a peripheral nerve causes paresthesia, hypalgesia, or hyperesthesia—not pain. After significant pressure, numbness and paresis result, but pain does not result. The conditions of nerve pressure from carpal tunnel syndrome, ulnar nerve palsy, and radial nerve palsy are examples of such situations.

Nerve tumors, whether they are benign or malignant, are not always painful. Neuromata are also frequently painless. Ischemia of a peripheral nerve has been considered to cause pain, but this concept is vague. The "sciatica" from nerve pressure of a herniated disk in the vertebral column is considered to result from pressure, irritation, or traction of the nerve's dural sheath and not from pressure on the nerve's axon.

Pain from soft tissue injury is transmitted by nociceptive substances created within the injured tissues; these substances lower the membrane potential at the end-organs of specific nerve fibers that ultimately transmit the stimulus to the pathways described earlier. This process is not "nerve pain."

Tendons

Tendons have universally been considered "inert" collagen fibers with no blood supply or innervation. Repair of injured tendons has always been considered to result from invasion of blood vessels from the peritenon tissue and from scar formation. There is now laboratory evidence that tendons intrinsically repair themselves by fibroblast stimulation, and that tension upon these fibroblasts realign them in a parallel manner, multiply them, and ultimately "repair" the damaged tendon. Tendons have been known to have an intrinsic nerve supply with unmyelinated nerves that carry proprioception, supply Golgi end-organ transmission, and transmit pain sensations.

All soft tissues, therefore, carry pain sensation and participate in articular impairment. Diagnostic examination is an attempt to discern the basis for the creation of the nociceptive process. The aim of the prescribed therapeutic regimen is to intervene in this nociceptive process within the soft tissues.

POSTURE

Posture will be mentioned frequently in discussions of the production of pain and functional impairment. There are many factors in the relationship of posture to pain that merit discussion.

The deep-seated patterns of emotionally evoked postures were well expounded by Feldenkrais,[11] who stated that improper head balance was rare in children except in those with structural abnormalities. He further claimed that emotional upheavals caused the child to assume postures that ensured safety. The flexed posture assumed and retained by the emotionally stressed child actively causes the flexor muscles to contract. The extensor muscles of the vertebrae thus become inhibited and have a gradual adverse influence on posture.

This flexed posture adversely influences walking and running as well as standing and sitting. It causes flexion of the hips and a forward vertebral dorsal (kyphotic) posture. The center of gravity intervenes, and a forward head posture results from increasing the cervical lordosis. The shoulder girdle also rotates to compensate. A thoracic deformity with subsequent respiratory diminution may result.

This assumed posture becomes habitual and "feels normal." Despite its "accepting" this posture, the body does not compensate, and the abnormal neuromuscular demands on the body that are required to maintain balance result in fatigue and mechanical stress on all elements of the musculoskeletal system. Pain and impairment result, and secondary structural changes occur.

METHOD OF EXAMINATION

To evaluate the patient presenting with potential soft tissue pain and functional impairment, and to determine the cause and site of pain, the examination must be very precise. The standard medical diagnostic format consists of a full history, a physical examination, and when indicated, confirmation by appropriate laboratory tests. The examination of the patient with soft tissue impairment involves the same general format, but with some modifications and refinements.

Examination of the patient impaired by or having pain from a soft tissue injury requires that the examining physician have a thorough knowledge of *functional* anatomy. The conclusion of the diagnostic examination must reveal *what* tissue has been impaired, *how* the precise tissue has sustained injury, and *what* needs to be done to correct or modify the impairment.

Two assumptions in the examination of the musculoskeletal system are (1) that the history reveals the exact mechanism by which a tissue has been impaired, and (2) that the movement and position causing the patient's pain can be ascertained when they reproduce, in part, the symptoms complained of by the patient. As all parts of the musculoskeletal system are capable of movement, and as all movements require the participation of the soft tissue, the claimed symptom is reproduced when the involved soft tissue is stressed.

The history reveals what the tissue insult is and how the person sustained it. The history also reveals the tissue site of the alleged painful insult and what portion of the functional anatomy was active when the insult occurred. The history should also reveal why the insult occurred. An activity may have been performed in an abnormal manner because of lack of precise training, fatigue, distraction by emotional factors, overuse, and so forth.

The history is probably the most important aspect of any examination and requires time, patience, effort, and discernment on the part of the examiner. The history may be voluntarily supplied by the patient, or revealed by actions, emphasis, innuendoes, tone of voice, or grimaces, but it must then be confirmed by an accurate examination.

The examination confirms what the history has implied. The history reveals the anatomic site of the insulted soft tissue. The musculoskeletal system is examined visually and by palpation. The involved part is actively, then passively, moved. The extent of movement is recorded, and the ensuing pain, discomfort, limitation, or abnormality of specific movement is noted. The manifested pain or discomfort is also noted as resulting from the imposed movement and as being appropriate and related to that movement or action.

The claimed limitation of action or movement must be directly attributed to the observed limitation. The exact soft tissue causing the pain and limitation is revealed from this examination, and this soft tissue can now be related to the activity that caused the insult, as claimed in the patient's history. In addition, the professed limitations can now be related to the involved offended tissues. *Reproduction of symptoms becomes meaningful, whether from the history or during active and passive examination activities.*

Trauma is the culprit of musculoskeletal pain and impairment, but is often difficult to define. External mechanical trauma can often be revealed by the history. Self-induced trauma is usually not that clear, and at times is not even recognized. Therefore, an accurate history must be elicited to determine the specificity and the magnitude of the self-induced trauma (Fig. 2–9).

Laboratory Confirmation

The third phase of a meaningful examination is laboratory confirmation. This confirmation involves x-ray studies, blood tests, and neurologic examinations such as electromyography (EMG) and nerve conduction times. More recently, sophisticated tests that reveal more soft tissue than previously have been discovered and perfected. These include computerized tomography (CT) scans, nuclear bone scans, and more recently, magnetic resonance imaging (MRI) tests. Also, the evoked cortical potential neurologic tests have further developed the previous (EMG) tests. All of these methods of testing will be discussed in relation to specific neuromusculoskeletal aspects of the body in subsequent chapters.

Unfortunately, the most sophisticated laboratory tests fail to reveal the exact status of soft tissue involvement. Nothing has yet replaced an accurate history and physical examination. As previously stated, abnormal laboratory tests must confirm what the history and physical examination have suggested; unfortunately, it is too often the abnormal laboratory finding that is treated rather than the patient presenting with the symptoms. Ordering then precisely interpreting the proper laboratory tests is the challenge to the physician.

DIAGNOSIS

Unfortunately, the **diagnosis**, the label of the offending tissue impairment, remains in a state of flux. **Taxonomy** is an evolving science and a needed aspect of medicine. The diagnosis too often relies on a label that is archaic, inaccurate, or meaningless. The involved tissue causing the pain and impairment is often not included in the diagnosis, nor is the mechanism of trauma clarified.

Terms such as arthritis, fibrositis, degeneration, tension state, "disk disease," and so forth are bandied about to the detriment of the patient, the legal profession, the third party carrier, and compensation awards. There is currently a sincere effort by the medical profession to clarify taxonomy and make "labels" for medical conditions descriptive and meaningful. The exact soft tissue involved must eventually be incorporated into this

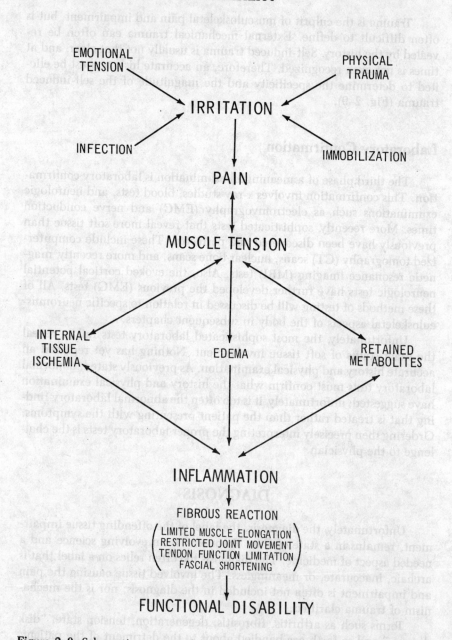

Figure 2-9. Schematic concept of functional disability related to soft tissue involvement.

taxonomy. Proper terms of impaired functional anatomy and the mechanisms by which the impairment occurred must evolve.

RATIONALE FOR SPECIFIC METHODS
OF TREATMENT

Assuming that there is validity to the neurologic and chemical components of pain production from soft tissue insult or injury, a rationale for the application of the various modalities must be established and verified. The primary objectives of treatment for soft tissue injuries are to minimize or remove the stimulus producing pain and to prevent or correct the tissue changes leading to impairment. A valid modality must be used to accomplish these objectives.

By adequately treating patients with acute soft tissue pain and impairment, the physician can help avoid or minimize progression into the recurrent or chronic stage.

Rest

Resting the injured part has endured as a method of treatment, but the implications of the term "rest" have not undergone careful scrutiny, nor has the beneficial duration of rest been defined. Currently, the duration of rest is being revised, and shorter periods are being advocated. This is so because the previously recommended times of inactivity have been found to be both physiologically and psychologically detrimental.

The recovery needs of tissue are being studied, the benefits of rest are being evaluated, and the benefits of activity (rather than inactivity) are being explored. The benefits of activity are becoming more evident than the previously claimed "benefits" of inactivity.

Rest was considered to allow natural recovery of the injured tissues. Inactivity has proven, however, to decrease the blood supply to and the removal of undesired metabolic waste products from the injured tissues. Weakness and atrophy of muscular tissue have been shown to occur sooner and to a greater degree than previously believed. Bone decalcification has been documented to occur rapidly. Psychologic dependency, debility, dependence on drugs, and despondency leading ultimately to depression have been found to occur very soon. All of these effects have been attributed to prolonged inactivity originally considered to be beneficial.

Detrimental effects of inactivity on specific tissues are being documented. The loss of muscle strength and endurance are well documented. Bone decalcification from inactivity and from loss of postural stimulus has been accepted since the early studies of patients with spinal injuries. Cartilage has been clearly shown to depend on compression—weight bearing and muscular contraction—for its normal nutrition.

Joint capsules have been found to require daily or at least frequent elongation to maintaian their flexibility. The cross linkage of collagen fibers composing the capsule has been shown to require frequent movement. Such movement prevents adherence of the intersecting fibers, which decreases the elasticity of the total capsule tissues.

Long immobilization of injured capsular tissues has been shown to cause deposits of calcium on the tissues. Edema formed at the site of injury causes deposits of protein outside the arterial capillary areas, causing a nidus for organization and fibrosis of the tissue space.

Pain has been considered a "reason" for resting the part. The reduction of pain resulting from rest has been documented and accepted. Frequently, however, a patient's pain is reduced from enforced activity rather than inactivity. Enforced inactivity leads to dependency, muscular disuse, drug use, and depression, all of which lead to chronic pain.

Minimizing the accumulation and the persistence of the nociceptive tissue stimulants at the site of the insulted soft tissue is the desired result of treatment. Resting the part for any length of time defeats this purpose.

Trauma to peripheral tissues is known to activate A gamma and C fiber endings by creating nociceptive chemicals as well as mechanical elements at these endings. The micro-environment of the end-organs consists of smooth muscles, blood capillaries, lymphatic vessels, and small unmyelinated nerves. There are also complex biochemicals that undergo pathophysiologic changes due to tissue injury. These have, in part, been identified as H^+ ions, serotonin, histamine, adrenaline-like substances, substance P, and kinins. Others have yet to be clarified or identified. All these substances have an excitatory effect on the membrane potential of the receptor end-organs. Impulses proceed centrally from the area of stimulation as stated previously in this chapter.

Ice

The local application of ice to the injured tissues has an accepted role in treating acute pain. Ice produces an analgesic effect, presumably by delaying the nerve conduction time of the nerves transmitting impulses. Its effect on the muscle spindle has been postulated to decrease muscle spasm.

In addition, ice is a vasoconstrictor and thus neutralizes the effects of histamine. Further edema and even microscopic hemorrhage are delayed and minimized. These effects are produced mechanically at the injured nerve site. Ice may also intervene in the conversion of arachidonic acid to prostaglandin B.

Sympathetic adrenergic substances at the periphery contribute to pain. These are produced from trauma-stimulated impulses transmitted via the sympathetic nerves. A persistent pain with its concomitant hypersensitivity can lead to reflex sympathetic dystrophic changes if this cycle is not interrupted early.

Ice also initiates impulses to the gray matter of the spinal cord. These impulses close the gate within the dorsal column (see Fig. 2–4).

Heat

Local application of heat to the inflamed tissues allegedly brings in fresh blood supply to remove the nociceptive elements. Repair of injured tissues is enhanced by fresh oxygen brought in after removal of the toxic elements.

A valuable aspect of heat treatment is that collagen fibers, which adhere from prolonged immobilization, tend to unlink with elevation of temperature. Capsular flexibility is thus regained. Early application of heat may even prevent adherence of collagen fibers.

Muscle accumulation of lactic acid, which causes pain and "spasm," is also reduced by vasodilatation.

Ice and heat are best considered as adjunctive therapy to allow early mobilization. To consider ice or heat in any form to be "definitive" therapy is an error. It is activity (exercise) that is ultimately beneficial, and activity is enhanced by the use of these modalities.

Exercise

In the long run, it is exercise that benefits injured soft tissues. All other modalities, such as heat, ultrasound treatment, hot packs, infrared treatment, or ice from any source, benefit the involved soft tissues merely by allowing them to function and thus minimizing pain, contracture, atrophy, and disuse.

Exercise is too vast a subject to receive more than brief mention at this point in the text. Specific exercises will be fully discussed in subsequent chapters in relation to specific extremities.

Exercise may be classified as active, passive, active-assisted, or active-resisted. Which type of exercise is prescribed depends on what the exercise is intended to accomplish.

Exercise may be used to regain or maintain flexibility; to maintain or regain strength or endurance; to realign the involved anatomic part; or to restore neurologic control, performance, or habit. After clarifying the soft tissue injury, the history and the physical examination will imply which type, frequency, and intensity of exercise are needed.

CONCEPTS OF CHRONIC PAIN

Chronic pain is considered to occur when there have been a persistence of acute pain and a failure of therapeutic response. The same concepts of acute pain and the same neurologic mechanisms that have been reviewed previously in this chapter are operational in chronic pain. These mechanisms persist, however, in transmitting impulses via the central neurologic patterns although the peripheral nociceptive stimuli may no longer be operating.

The soft tissue nociceptive stimuli may have vanished, been diminished, or been changed; or the stimuli may still be operational, but the central response may no longer be compatible with these peripheral stimuli. The "concept" of pain is now deep-seated and recalcitrant to any peripheral intervention.

Originally, a period of 6 months was arbitrarily considered the time when acute pain becomes chronic pain. Gradually, this period was shortened, and now there are some scientists and clinicians who maintain that chronicity becomes apparent within three to six weeks.

These same authorities also claim that the term chronic pain implies that all acceptable therapeutic approaches have failed to relieve the symptoms and that the patient becomes or remains disabled far in excess of the apparent diagnosed impairment. The cause of the pain now becomes obscure, and the symptoms become relatively inexplicable.

"Primary gains" denote justification of rest, and assistance, support, and concern from the spouse and others. These should be commensurate with the extent of injury. "Secondary gains" are monetary, or assistance and concern from spouse and others for exceeding the extent of injury.

The "secondary gains" of the disability now are assumed to play a major role. These gains may be physical, emotional, psychologic, psychosocial, social, or even monetary.

In chronic pain, the term pain describing a "symptom" is now considered to describe a "disease entity."

Two major schools of thought regarding the concept of "chronic pain" predominate. These schools can simplistically be termed centralist and peripheralist.

The peripheralists consider that chronicity intervenes because the peripheral nociceptive stimuli continue to bombard the dorsal columns and maintain the pain-producing pathways.

The centralists consider that the bombardment has initiated a self-reverberating, self-perpetuating pattern that no longer needs peripheral stimuli. All pain is "central" from the dorsal columns to the ascending pathways, the limbic system, and significantly, the cerebral cortex. Among the centralists are also the psychologically oriented physicians who consider the psychologic aspect as predominant if not the exclusive causative factor maintaining the disability.

The neural mechanisms of chronic pain remain in the evolutionary phase. The centralists believe that the reverberating patterns exist in the Rexed layers of the dorsal horn and are influenced by the descending impulses from the limbic and raphe system.

The psychophysiologists feel that chronic pain is an emotional burden on the psyche of the sufferer and that psychotherapy is the only recourse for recovery.

With so many varied schools of thought, there are an equal number of differing advocated treatments. It is probable that an aspect of all schools plays a part in the production and maintenance of chronic pain. Treatment today uses an aspect of each in efforts to alleviate chronic pain.

It stands to reason that correcting and alleviating the acute pain *early and completely* constitute a reasonable approach to the prevention of chronic pain. Recognizing the overwhelming injury and the severe soft tissue insult with acceptable limited recovery assures that the injured person receives appropriate therapy for *all* aspects of the pain.

Early recognition of injured people who are likely to develop chronic pain from tissue pathology also prevents unecessary, useless, expensive, and frustrating treatments. Patients whose psychologic makeup, regardless of the psychologic school of thought, is such that they "need" their symptom and essentially "lie"[12] to themselves usually fail to recover from soft tissue injury and refute all the acceptable physiologic aspects of soft tissue pain and disability.

Recognizing the significance of acute soft tissue injuries and apprising patients of the problems to enlist their assistance in the therapeutic venture are major factors in minimizing the frequency and severity of chronic pain.

Recognition, evaluation, and treatment of the chronic pain patient in today's medical environment remain the most perplexing problems of the practitioner.

REFERENCES

1. Steinbach, RA, et al: Chronic low back pain: "Low-back loser." Postgrad Med 53:135–138, 1973.
2. Melzak, R, and Wall, PD: Pain mechanisms—a new theory. Science 150:971, 1965.
4. Froriep: Ein Beitrag zur Pathologic and therapie des rheumatesmus. Weimar, Germany, 1843.
5. Grauhan, M: Überden anatomeschen Befience bel einem Fall von myositis Rheumatica. Doctoral dissertation. Cassel, Weber, & Weidemeyer, 1912.
6. Steindler, A, and Luck, JV: Differential diagnoses of pain in the low back. JAMA 110:106–113, 1938.
7. Travell, J, et al: Pain and disability of shoulder and arm: Treatment by intramuscular infiltration with procaine hydrochloride. JAMA 120:417, 1942.

8. Simon, DG: Muscle pain syndromes. Special review. Am J Phys Med, Williams & Wilkins, Baltimore, Vol 54, No 6, 1975, Part I; Vol 55, No 1, 1976, Part II.
9. Travell, J: Ethyl chloride spray for painful muscle spasm. Arch Phys Med 33:291-298, 1952.
10. Kraft, JH, Johnson, EW, and LeBan, MN: The fibrositis syndrome. Arch Phys Med 49:155-162, 1968.
11. Feldenkrais, M: Body and Mature Behavior, ed 3. International Universities Press, New York, 1975.
12. Peck, MS: People of the Lie. Simon & Schuster, New York, 1983.

BIBLIOGRAPHY

Altschule, MD: Emotion and skeletal muscle function. Med Sci 11:163-164, 1962.
Anrep, GJ, Saalfeld, EV: The blood flow through the skeletal muscle in relation to its contraction. J Physiol 85:375-399, 1935.
Awad, EA: Interstitial myofibrositis: Hypotheses of the mechanism. Arch Phys Med Rehabil 54:449-453, 1973.
Barcroft, H, Millen, JLE: The blood flow through muscle during contraction. J Physiol 107:518-526, 1948.
Basbaum, AI, and Fields, HL: Endogenous pain control systems: Brainstem spinal pathways and endorphin circuitry. Ann Review Neuroscience, 7:309-338, 1984.
Berges, PV: Myofascial pain syndromes. Postgrad Med 53:161-168, 1973.
Bonica, JJ: The Management of Pain. Lea & Febiger, Philadelphia, 1953.
Bowsher, D: Termination of the central pain pathways in man: The conscious appreciation of pain. Brain 80:606-621, 1957.
Chapman, CR, and Turner, JA: Psychological control of acute pain in medical setting. J Pain and Symptom Management, 1(1):9-20, Winter 1986.
Clark, WC, and Hunt, HF: Pain. In Downey, JA, and Darling, RC (eds): Physiological Bases of Rehabilitative Medicine. WB Saunders, Philadelphia, 1971, pp 373-401.
Cobb, CR, et al: Electrical activity in muscle pain. Am J Phys Med 54:80-87, 1975.
Cure, BL: Pain. The Bull, Los Angeles Co Med Assoc, Oct 4, 1973, pp 10-15.
Crue, BL, and Carregal, EJA: Post synaptic repetitive neurone discharge in neuralgic pain. Presented at International Symposium on Pain, May 1975, Seattle. Raven Press, New York, 1976.
deVries, HA: Quantitative electromyographic investigation of the spasm theory of muscle pain. Am J Phys Med, 45(3):119-134, 1966.
Edmeads, J: The physiology of pain: A review. Prog Neuropsychopharmacol Biol Psychiatr 7:413-419, 1983.
Feinstein, B, et al: Experiments of pain referred from deep somatic tissues. J Bone Joint Surg 36[Am]:981-997, 1954.
Fordyce, WE: Operant conditioning as a treatment method in management of selected chronic pain problems. Northwest Med 69:580, 1970.
Forst, JJ: Contribution a l'etude clinique de la sciatique. Paris These, No 33, 1881.
Gordon, G: Role Theory and Illness: A Sociological Perspective. College & University Press, New Haven, CT, 1966.
Hirschfeld, AH, and Behan, RC: The accident process. JAMA 186:193-199, 1963.
Holmes, T, and Rahe, R: The social readjustment rating scale. J Psychosom Res 11:213-218, 1967.
Holmes, TH, and Wolff, HG: Life situations, emotions, and backache. Psychosomat Med 14:18, 1952.

Jacobson, E: Electrical measurements of neuromuscular states during mental activities. I. Imagination of movement involving skeletal muscles. Am J Physiol 91:567–608, 1930.

Kerr, FW: Pain: A central inhibitory balance theory. Mayo Clin Proc 50:685–690, 1975.

King, JS, and Lagger, R: Sciatica viewed as a referred pain syndrome. Surg Neurol 5:46–50, 1976.

Lasegue, CH: Considerations sur la sciatique. Arch Gen Med 2 (Serie 6, Tome 4):558–580, 1864.

Layzer, RB, and Rowland, LD: Cramps. N Engl J Med 285:31–40, 1974.

Livingston, WK: Pain Mechanisms. Macmillan, New York, 1943, pp 128, 139.

Lundervold, A: Electromyographic investigation during sedentary work. Br J Phys Med 14:32–36, 1951.

Maigne, R: Medical Orthopedics. Charles C Thomas, Springfield, IL, 1975.

Malmo, RB, and Shagass, C: Psychosomat Med 11:9, 1949, and 11:25, 1949.

Mead, GH: Mind, Self and Society. University of Chicago Press, Chicago, 1952.

Mehler, WR: Some observations on secondary ascending afferent systems in the central nervous system in pain. Henry Ford Hospital International Symposia, Little Brown & Co, Baso Boston, 1964, pp 11–32.

Melzack, R: Myofascial trigger points: Relation to acupuncture and mechanisms of pain. Arch Phys Med Rehab 62:114–117, 1981.

Melzak, R, and Wall, PD: Pain mechanisms—a new theory. Science 150:971, 1965.

Melzack, R, Stillwell, DM, Fox, EJ: Trigger points and acupuncture points for pain: Correlations and implications. Pain 3:3–23, 1977.

Miglietta, O: Action of cold on spasticity. Am J Phys Med vol 52, 52(4):198–204, 1973.

Minter, WJ, and Barr, JS: Rupture of the intervertebral disk with involvement of the spinal canal. N Engl J Med 211:210, 215, 1934.

Nathan, PW: The gate-control theory of pain: A critical review. Brain 99:123–158, 1976.

Neufeld, I: Mechanical factors in the pathogenesis, prophylaxis and management of "fibrositis" (fibropathic syndromes). Arch Phys Med Rehabil 759–765, 1955.

Park, SR, and Rodbard, S: Effects of load and duration of tension on pain induced by muscular contraction. Am J Physiol 203:735–738, 1962.

Parsons, T: The Social System. Free Press, Glencol, IL, 1951.

Perl, ER: Mode of action of nociceptors. In Hirsch, C, and Zotterman, Y (eds): Cervical Pain. Pergamon Press, Oxford, 1972.

Rees, WS: Multiple bilateral subcutaneous rhizolysis of segmental nerves in the treatment of the intervertebral syndrome. Ann Gen Pract 26:126–127, 1971.

Sainsbury, P, and Gibson, JG: Symptoms of anxiety and tension and the accompanying physiological changes in the muscular system. J Neurol Neurosurg Psychiatry 17:216–224, 1954.

Shealy, CN: Facets in back and sciatic pain: A new approach to a major pain syndrome. Minn Med 57:199–203, 1974.

Simons, DG: Muscle pain syndromes, Part I. Am J Phys Med 54:289–311, 1975.

Stillwell, DL: The innervation of tendons and aponeurosis. Am J Anat 100:289–317, 1957.

Travell, J: Pain mechanisms in connective tissue. In Regan (ed): Conference on Connective Tissue, Josiah Macy Jr Foundation, New York, 1951, pp 86–125.

Travell, J, and Rinzler, SH: The myofascial genesis of pain. Postgrad Med 11:425–434, 1952.

Walter, A: The psychogenic regional pain syndrome and its diagnosis. In: Pain. Little Brown & Co, Boston, 1966, Ch 34.

CHAPTER 3

Low Back Pain

Low back pain constitutes the major musculoskeletal disabling condition in modern society. It has been estimated that 80 percent of the general population will have a complaint of low back pain and disability in the course of their lives. Recent surveys have stated that approximately 550 million days of work are lost annually as a result of pain and that low back pain constitutes 56 percent of this pain complaint. The loss of work time has been estimated to cost the American economy 55 billion dollars annually. Of patients suffering from low backache, 42 percent have consulted three or more doctors for this condition.[1]

The assumed causes of low back pain are myriad. The fact that there are numerous diagnostic labels assigned to the symptoms of mechanical low back pain implies the lack of universal agreement as to the exact causation.

Organic causes such as fractures, malignancy (either primary or metastatic), osteomyelitis, diskitis, or referred pain from a diseased visceral organ must be suspected, and appropriate tests must be conducted, if the history and examination fail to indicate a purely mechanical cause. Once these organic causes are ruled out, appropriate neuromuscular, skeletal, and psychologic studies must be performed to establish the cause. Treatment must then evolve based on the etiologic conclusion.

Mechanical low back pain can be simplistically designated as acute, recurrent, or chronic. Low back pain can be further categorized as either static or kinetic. These arbitrary designations are reached by appropriate history, meaningful examination, and significant laboratory tests.

The functional anatomic basis of pain and disability in the lower back is becoming increasingly understood. The tissue site of nociceptive pain initiation is more clearly recognized, as is the deviation from normal functional anatomic factors that initiated the insult or injury. Determining the deviation from normal function is the basis for performing diagnostic procedures and for planning appropriate treatment.

VERTEBRAL COLUMN

The spine of man functions as a total of superincumbent functional units. Each unit functions uniquely but in concert with the adjacent units (Fig. 3–1). Each component part of the functional unit has a precise function and may also be the site of malfunction and pain stimulus. The total spine, five functional units in the lumbosacral segment, functions in a precise mechanical manner (Fig. 3–2).

PHYSIOLOGIC CURVES

There are four physiologic curves when the total spine is viewed laterally: the cervical curve, a lordosis, composed of eight vertebrae; the dorsal curve, a kyphosis, composed of eight vertebrae; a lumbar curve, a lordosis, formed by five lumbar vertebrae; and a sacral curve, composed of five fused sacral vertebrae.

All four physiologic curves are formed by the superincumbent vertebral functional units, which are precariously balanced upon the sacrum (Fig 3–3). Each curve adjoins its immediate adjacent curve with a transitional region suited for a precise function. Each functional unit of each curve has a possible movement that is unique for that functional unit, and thus, for that segment of the vertebral column.

These physiologic curves are not present at birth (Fig 3–4); they develop as the person develops. The newborn spine is a total kyphotic curve formed by the shape of the vertebral bodies and their intervertebral disks. As the child matures, the lordotic curves develop. At first, the head is elevated with the development of the cervical lordosis. Gradually, as the erect posture is assumed, the lumbar spine develops a lordosis. The lordosis formed in the lumbar region is assumed to be the result of the spine extending but being resisted by the iliopsoas, which fails to elongate adequately.

By the time the child's posture becomes totally erect, the vertebral column has acquired four physiologic curves supported on the sacral base. These are the cervical and lumbar lordotic and dorsal and sacral kyphotic physiologic curves. They are balanced upon an oblique sacral base (Fig. 3–5).

The balance upon the sacral base depends on the angulation of the sacrum, which in turn is related to the pelvis as it articulates upon the two

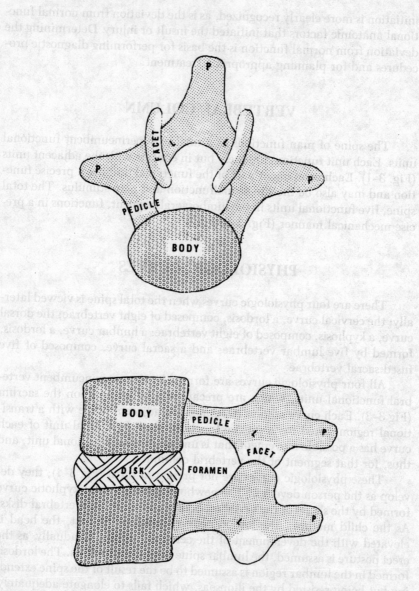

Figure 3–1. Functional vertebral unit. *Top,* View of the vertebral body, the posterior articulations (facets), the pedicles, the processes (P), and the lamina (L). *Bottom,* Lateral view demonstrating the intervertebral disk and its relationship to the components of the unit.

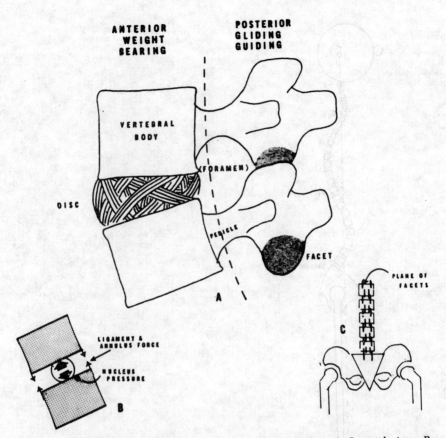

Figure 3–2. The functional unit of the spine in cross section. *A*, Lateral view. *B*, Diagram shows pressure within disk, which forces vertebrae apart, and the balancing force of the long ligament. *C*, Gliding motion of the plane of the facets.

hip joints. Increased angulation of the sacral base increases the lordosis of the superincumbent immediate spinal curve.

ERECT BALANCE AND STANCE

Erect balance is a biomechanical function. As each functional unit of the vertebral column is flexible, each spinal curve is also flexible. Balance upon the sacral base requires that the base is stable and that each curve remains as close as possible to the center of gravity.

The spine must stand erect with maximal efficiency and minimal expenditure of energy. The use of muscular contraction or "tone" alone to support the spine with its physiologic curves being away from the center of

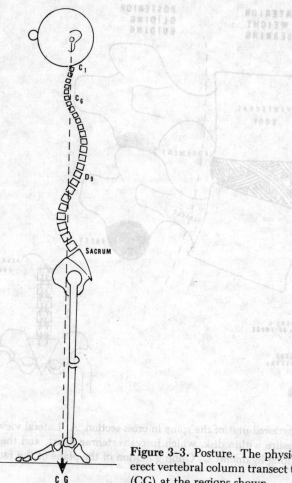

Figure 3-3. Posture. The physiologic curves of the erect vertebral column transect the center of gravity (CG) at the regions shown.

gravity is an impossibility. Therefore, erect support must derive from ligaments and from pressure within the disks, so that minimal effort and expenditure of energy are required.

During standing or sitting, each functional unit plays a vital role in weight bearing. In bending, stooping, twisting, turning, and lifting, each functional unit plays a proportional role in the total movement.

As well as mechanically supporting the body, each functional unit contains a bony canal that encloses and protects the neural contents of the cord and the cauda equina. These bony posterior structures also act as points of attachment of the ligaments and the muscles that support the erect spine and cause it to function kinetically.

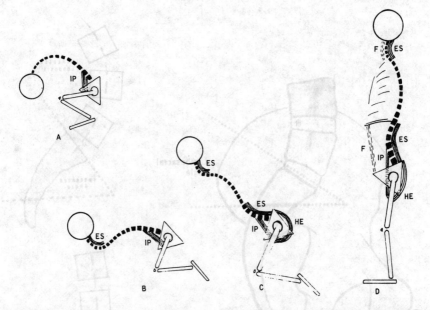

Figure 3-4. Evolution of erect posture. *A,* Fetal position with total kyphosis; *B,* Extension of cervical spine with elevation of head; *C,* Extension of lower extremities beginning stretch of iliopsoas muscle, causing lumbar lordosis; *D,* Fully erect spine with cervical lumbar lordosis and dorsal and sacral kyphosis.

F—flexors
ES—extensors
IP—iliopsoas
HE—hip extensors

THE FUNCTIONAL UNIT

The "functional unit," the basic structural component of the spine, is composed of two adjacent vertebrae (see Fig. 3–1). The functional unit is composed of two segments: the anterior weight-bearing portion and the posterior neural canal, which contains and protects the neural structures and simultaneously provides points of attachment for the supporting and operative tissues of the spine. In the posterior segment, there are also structures that guide and direct the movement of the functional unit.

Anterior Portion

The anterior portion of the functional unit is constructed so that it is weight bearing and permits movement. It comprises two adjacent verte-

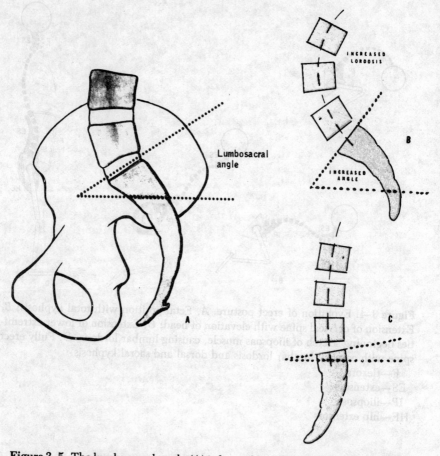

Figure 3–5. The lumbosacral angle (*A*) is formed by a line paralleling the top of the sacrum and a line drawn horizontally. The spine is balanced upon the sacrum. As the angle increases (*B*), the lordosis is greater. With a lesser angle (*C*), the lumbar lordosis is less.

bral bodies separated by the intervertebral disk (see Fig. 3–2). The vertebral bodies are rounded bones with flattened superior and inferior surfaces. These surfaces are covered around the periphery by hyaline cartilages. The epiphyses of the vertebral bodies close at puberty, and except for severe external trauma or metabolic deterioration, remain intact throughout life.

The disk is a fibrocartilaginous tissue comprising the following:

1. The annulus fibrosus, which is composed of concentric lammellae of collagen fibers that attach from the periphery of the vertebral

bodies and obliquely proceed to attach to the adjacent vertebrae (Fig. 3–6). These collagen fibers are enmeshed within a mucopolysaccharide matrix, which contains approximately 88 percent water at birth. This percentage decreases gradually and ranges from 70 to 78 percent at senescence.

2. The nucleus pulposus, a centrally located mass of mucopolysaccharide contained within the annular envelope and between the adjacent vertebral end-plates.

The disk functions hydrodynamically. The pressure within the nucleus is exerted in all directions and thus separates the vertebral end-plates. The annular fibers have sufficient flexibility to allow change in the configuration of the nucleus (Fig. 3–7).

The annular fibers function similarly to their component collagen fibers. Collagen fibers, which are tri-helical chains of amino acids, have the ability to extend (elongate) by completely uncurling. They also have the characteristic of resuming their length after release of tension. These fibers are in a sheet encircling the entire disk. The outer sheet of parallel collagen fibers obliquely goes from the superior vertebral end-plate to the inferior end-plate at a 30-degree angle. The next inner sheet has the fibers running

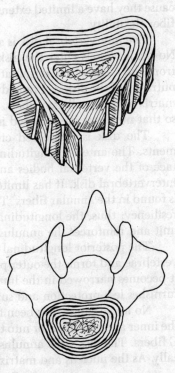

Figure 3–6. Annulus fibrosus. *Top,* Layer concept of annulus fibrosus. *Bottom,* Circumferential annular fibers about the centrally located pulpy nucleus (nucleus pulposus).

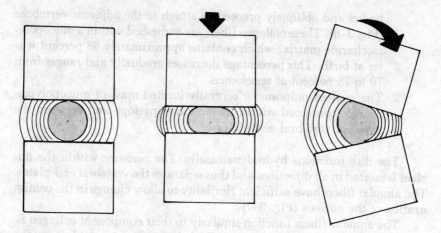

Figure 3-7. *Left,* Normal non-weight-bearing disk. *Center,* Deformation of nucleus reacting to compression. *Right,* Deformation of nucleus permitting flexion or extension.

in the opposite direction. These annular structures have flexibility because they are in sheets, because they intertwine in opposite directions, and because they have a limited extensibility by virtue of the component collagen fiber flexibility.

The intervertebral disk is an avascular tissue that has no nerve supply. No blood vessels enter the disk after puberty. Blood approaches the disk from the vessels that terminate at the end-plates of the vertebrae. The nutrition of the disk is obtained from this blood supply by imbibition of the matrix and the collagen fibers. Compression and release of the disk occur so that nutrients are imbibed mechanically and chemically.

The disk is reinforced circumferentially by the longitudinal ligaments. The anterior longitudinal ligament covers the entire anterior surface of the vertebral bodies and becomes the anterior outer layer of the intervertebral disk. It has limited flexibility and a stronger resiliency than is found in the annular fibers. The parallel alignment of the fibers also add resiliency; thus, the longitudinal ligament limits motion of the functional unit and reinforces the annulus of the disk.

The posterior longitudinal ligament covers the posterior aspect of the vertebrae and forms the outer posterior layer of the intervertebral disk. As it becomes narrowed in the lumbosacral region of the spine (Fig. 3–8), it furnishes less protection and support at this level.

No nerve supply has been traced into the disk substance, that is, into the inner annular fibers or into the nucleus: mechanoreceptors, A fibers, or C fibers. The disk—the annulus and nucleus—acts entirely hydrodynamically. As the nucleus and matrix imbibe fluid, the disk expands and exerts

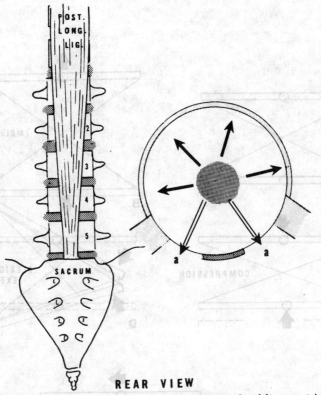

REAR VIEW

Figure 3-8. *Left,* Inadequacy of the posterior longitudinal ligament in the lower lumbar segment, which decreases the protective effect in the L_4, L_5, and S_1 region. *Right,* Disk herniation may bulge into the spinal canal, a.

pressure against the vertebral end-plates to separate them, and presses outward to extend the surrounding annular fibers. The hydrodynamic pressure supports the functional units and thus supports the vertebral column.

The hydrodynamic function permits weight bearing to occur, with deformation of the nucleus and some elongation of the surrounding annular fibers. Upon release of weight bearing, the disk resumes its original structure. With this compressibility, flexion or extension of the functional unit is also permitted (Fig. 3–9).

The flexibility of the annulus is permitted by (1) the angulation of the fibers forming the annular sheets, (2) the elongation of the uncoiling annular fibers, and (3) the lubrication between the annular sheets of collagen. The limitation of flexibility is the limited extension of the collagen fibers before they "tear" by unlinking the amino acid bonds (see Chapter 1).

Nutrition of the disk by imbibition occurs via transfer of nutrients through the end-plates and through the long ligaments. Solutes containing

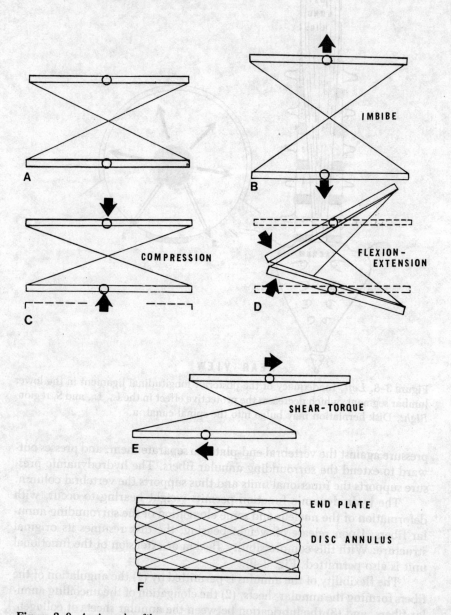

Figure 3–9. *A*, Annular fibers at rest. *B*, Effect of elongation. *C*, Effect of compression. *D*, Effect of flexion or extension. *E*, Effect of translatory torque.

glucose and oxygen enter through the end-plates, and sulfates are thought to enter via the long ligaments and the annular fibers (Fig. 3–10).

The annulus maintains its integrity so long as the external forces do not interrupt or damage the component parts. The collagen fibers of the annular sheets have sufficient elongation to permit bulging of the nucleus from compression and from limited rotation and shear. Too much rotation causes excessive elongation of the annular collagen fibers and can disrupt the amino acid chains, causing "rupture" (Fig. 3–11).

Excessive compression of a disk results in fracture of the bony end-plates before there is tearing of the annular collagen fibers. Excessive flexion and extension is resisted by the longitudinal ligaments before there is disruption of the annular fibers. Without intrinsic abnormality of the nucleus, the matrix, or the collagen fibers, *only* rotatory torque can disrupt the integrity of the disk. (see Fig. 3–11).

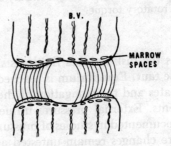

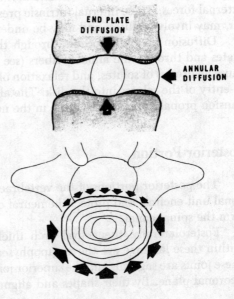

Figure 3-10. Disk nutrition through diffusion. Diffusion of solutes occurs through the central portion of the end-plates and through the annulus. Marrow spaces exist between circulation and hyaline cartilage and are more numerous in the annulus than in the nucleus. Glucose and oxygen enter via the end-plates. Sulfates enter through the annulus to form glucosaminoglycans. There is less diffusion into the posterior annulus. (B.V.—blood vessels)

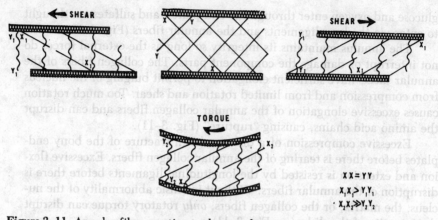

Figure 3–11. Annular fiber reaction to shear and torque. *Top left*, Shear to the left elongates X fibers, thus permitting Y fibers to undulate and shorten. *Top middle*, Normal positions. *Top right*, Contrary shear effect. *Below*, Effect on annular fibers in rotatory torque.

As the disk imbibes fluid, its size increases, and the annular fibers become taut. Equilibrium is reached between the two adjacent vertebral end-plates and the elongation of the annular fibers and the longitudinal ligaments. External pressures influence this equilibrium and have been well documented in studies of pressure within the disks. Although intrinsic pressure changes remain untested and unconfirmed, these changes may help to explain disk degeneration and herniation when combined with the external forces. These internal intrinsic pressure gradients may be molecular, may involve solutes, and may be under hormonal influence.

Diffusion of solutes occurs through the central portion of the end-plates and through the annular fibers (see Fig. 3–10). External pressures cause an exodus of solutes, and relaxation of the intervertebral disks causes re-entry of the solutes into the disk. The alternating compression and expansion probably play a vital role in the nutrition of the disk.

Posterior Portion

The posterior portions of the vertebrae that combine to form a functional unit encircle and protect the neural components of the spine. They form the spinal canal.

Posterolaterally, the neural arch thickens and forms the laminae. Within these laminae are the zygapophyseal ("facet") joints (Fig. 3–12). These joints are sagittal, and the superior joint opposes the inferior joint in a coronal plane. By their shapes and alignment, they articulate to allow

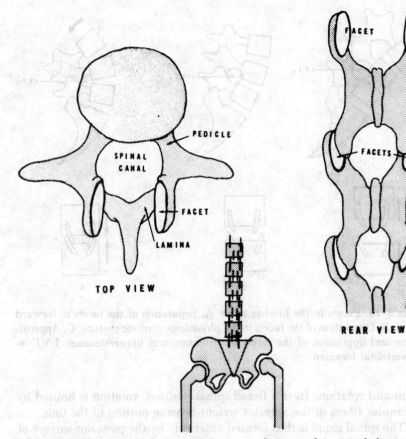

PEDICLE

SPINAL
CANAL

FACET

LAMINA

TOP VIEW

FACET

FACETS

REAR VIEW

Figure 3-12. Posterior view of the lumbar spine depicting the vertical alignment of the posterior articulation (facets) and depicting the small musculature of the lumbar column.

flexion and extension of the functional unit and to restrict, even prevent, lateral flexion and rotation of the functional unit. This articular relationship explains that motion of the lumbar functional units is essentially flexion and extension and explains why this restriction prevents torque stresses on the annular fibers of the intervertebral disk in spinal motions (Fig. 3–13).

These facets are coated with articular cartilage and have a capsule and synovial lining. They articulate upon each other but bear little or no weight. It has been estimated that in the full lumbar extended posture, they are capable of as much as 25 percent of weight bearing.

In the extended lumbar position, the facets are so opposed that they are presumably "locked," thus preventing any lateral-rotatory movement. In the flexed position, they separate and allow a few degrees of lateral

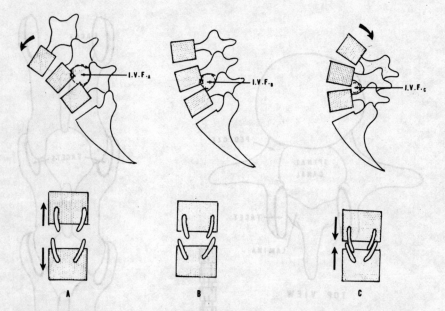

Figure 3-13. Facets in the lumbar spine. *A*, Separation of the facets in forward flexion. *B*, Opposition of the facets in the physiologic lordotic posture. *C*, Approximation and opposition of the facets on extension and hyperextension. I.V.F. = intervertebral foramen.

flexion and rotation. In this flexed spinal position, rotation is limited by the annular fibers of the anterior weight-bearing portion of the unit.

The spinal canal is thus formed anteriorly by the posterior surface of two adjacent vertebrae and the posterior longitudinal ligament lining the intervertebral disk, and posteriorly by the facets and their capsules. Also, the spinal canal is coated by the ligamentum flavum, an elastic ligament. The ligamentum flavum extends to the entire length of the spinal canal. Its primary function is considered to be prevention of constriction of the capsule of the facets between the articular surfaces during flexion and extension.

The bony posterior arch elongates posteriorly to form the posterior superior spine, and laterally to form the transverse processes. These are the points of attachment of the intervertebral ligaments and the posterior erector spinae muscles.

INNERVATION OF THE LUMBOSACRAL SPINE

Within the spinal canal resides the termination of the spinal cord, the cauda equina. This transition from cord to cauda equina occurs at the first lumbar-thoracic articulation. As the "roots" of the cauda descend, they gradually progress laterally, leaving the spinal canal and descending the lower extremities as sciatic and femoral nerve bundles (Fig. 3–14).

These nerve roots leave the spinal canal via the intervertebral foramina. Because of their vertical alignment, they course obliquely across the vertebrae, pass under the pedicle then across the posterior lateral aspect of the disk (Fig. 3–15).

Of clinical significance is the fact that the nerve roots are invested within a dural sac that contains spinal fluid and numerous arterioles, venules, lymphatic vessels, and nerve fibers (Fig. 3–16). Within this sheath are the motor and sensory fibers that form the peripheral nerves (Fig. 3–17). Each nerve root within its sheath passing through the foramen has a sensory and a motor fiber component as well as a sympathetic nerve fiber component.

The arachnoid that encloses the nerve roots continues along the nerve as far as the posterior sensory nerve root ganglion but does not encircle it completely. The accompanying dura continues along the sensory and motor nerve roots until it merges into one dural sheath (see Fig. 3–16). This dural sheath continues along the nerve root to form the perineurium.

The posterior sensory nerve root is twice the size of the motor root (see Fig. 3–17). The two roots remain separated until the sensory root expands into the ganglion. Within the foramen, the motor root is closest to the disk, and the sensory root is closest to the facet joints.

The roots and their investing sheaths are not attached to the walls of the intervertebral foramina; thus, some movement within the foramina is possible. There are, however, some connecting collagen fibers that attach the roots to the walls of the foramina, so that some traction forces upon the root are possible. The nerves and their investing sheaths occupy approximately 50 percent of each foramen. The remainder of the foramen is composed of adipose tissue and loose areolar connective tissue.

The cone-shaped structure of the dural sheath that accompanies the nerve roots in their passage through the foramina is considered to prevent traction injuries (Fig. 3–18).

TISSUE SITE OF NOCICEPTIVE PAIN PRODUCTION IN THE FUNCTIONAL UNIT

Pain produced from abnormality within the functional unit implies injury to a tissue within the unit. Irritation resulting in pain also implies a nerve ending within the tissue. The tissue site of pain production within the functional unit has been clinically ascertained and anatomically confirmed.

1. The posterior longitudinal ligament is innervated by the recurrent

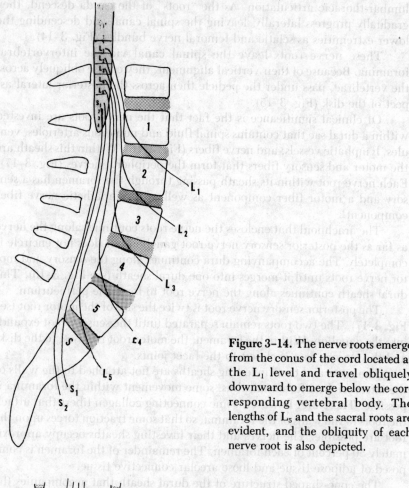

Figure 3–14. The nerve roots emerge from the conus of the cord located at the L_1 level and travel obliquely downward to emerge below the corresponding vertebral body. The lengths of L_5 and the sacral roots are evident, and the obliquity of each nerve root is also depicted.

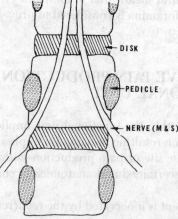

Figure 3–15. Relationship of nerve root to pedicles. The nerve descends the spinal canal and moves obliquely out, passing the pedicle and crossing the inferior disk at its posterior lateral aspect. There is little exposure of the nerve to the disk at this level. Most of the exposure is in the immediate cephalad disk. M & S = motor and sensory.

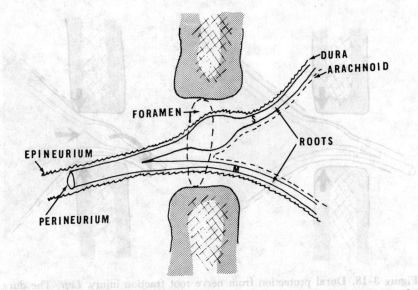

Figure 3–16. Dural and arachnoid sheaths of the nerve root complex. The arachnoid follows the sensory and motor nerve roots to the beginning of the intervertebral foramen and follows the sensory root to the beginning of the ganglion. The dura follows the nerve roots until they become the combined sensory and motor nerve outside the foramen, and it continues as the perineurium and epineurium. Neither the dura nor the arachnoid attach to the intervertebral foramen.

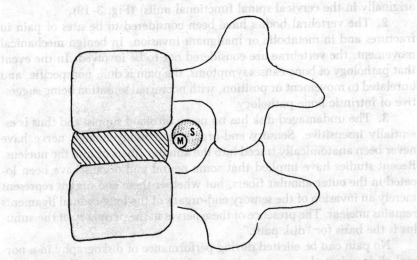

Figure 3–17. Nerve root complex emerging through foramen. When the sensory and motor nerve roots combine and emerge through the foramen, the motor fibers (M) are concentrated in the anteroinferior portion of the nerve root complex. (S—sensory fibers)

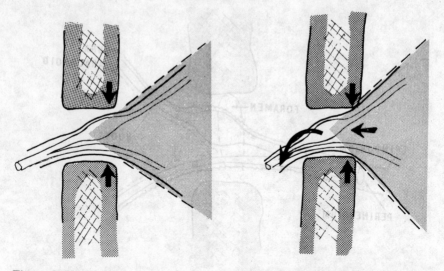

Figure 3–18. Dural protection from nerve root traction injury. *Left,* The dura forms a funnel, which keeps the nerve root complex free within the foramen. *Right,* Effective traction draws the nerve root and its dura into the foramen, where it becomes blocked mechanically at the site of the *arrows,* making further traction impossible.

nerve of Luschka and has been shown to cause pain when inflamed (this originally in the cervical spinal functional units) (Fig. 3–19).

2. The vertebral bodies have been considered to be sites of pain in fractures and in metabolic or malignant invasion. In benign mechanical movement, the vertebrae are considered not to be involved. In the event that pathology of bone causes symptoms, the pain is dull, nonspecific, and unrelated to movement or position, with nocturnal sensation being suggestive of intrinsic bone pathology.

3. The undamaged disk has no nerve or blood supply and thus is essentially insensitive. Sensory end-organs and unmyelinated nerve have never been anatomically traced into the annular fibers or into the nucleus. Recent studies have implied that some neural end-organs have been located in the outer annular fibers, but whether these end-organs represent merely an invasion of the sensory end-organs of the longitudinal ligaments remains unclear. The presence of these nerves in the periphery of the annulus is the basis for "disk pain."

No pain can be elicited during performance of diskography in a normal disk unless the pressure is excessive, as was originally stated by Hirsch.[2] Local anesthetic injection into the disk nucleus has not proven to be effective in treating low back pain. Injection into an abnormal disk does cause pain, but its neural mechanism remains unclear.

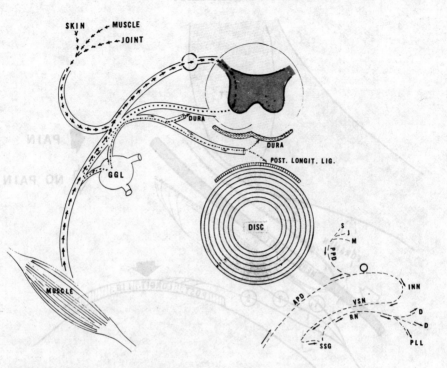

Figure 3-19. Innervation of the recurrent nerve of Luschka.

PPD—posterior primary division
APD—anterior primary division
GGL—sympathetic ganglion
INN—internuncial neurons
VSN—ventral sensory nerve
SSG—sensory sympathetic ganglion
RN—recurrent nerve of Luschka
D—to dura
PLL—posterior longitudinal ligament

4. The posterior longitudinal ligament is copiously supplied with end-organs of unmyelinated and myelinated nerves and is richly supplied by the recurrent nerve of Luschka. Clinically, this long ligament has been shown to be a site of low back pain when irritated chemically or mechanically.

5. The sheaths of the nerve roots within the foramina are amply supplied by branches of the recurrent nerve of Luschka and have clinically been confirmed as sites of back and referred leg pain. The recurrent Luschka nerve returns via the foramina and supplies the anterior and the posterior longitudinal ligaments and the nerve sheath dura (Fig. 3-20).

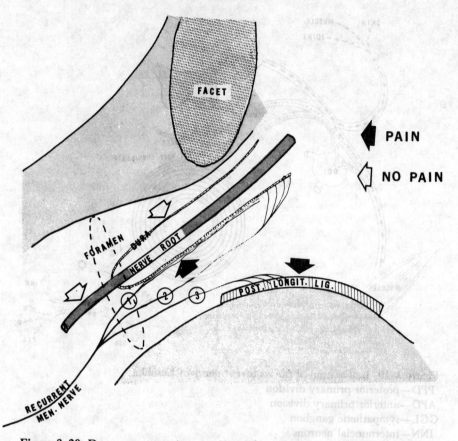

Figure 3-20. Dura accompanying nerve root through intervertebral foramen with its innervation by the recurrent meningeal nerve. Branches 1 and 2 proceed to the anterior dural sheath, illustrating the sensitivity of that portion of the dura. The posterior dural sheath with no innervation is insensitive. Innervation (3) of the posterolongitudinal ligament, which is capable of pain.

The innervation of the dura is found primarily on the dorsum of the nerve, and none is found on the ventral surface, which explains the type of pain elicited by various sources of nerve root pressure (see Fig. 3-20).

6. The yellow ligamentum flavum is totally without nerve supply and is therefore insensitive.

7. The interspinous ligaments are well innervated by neurons capable of transmitting nociceptive stimuli. Kellgren[3] caused radiating pain in the distribution of the sciatic nerve by injecting irritating substance into the interspinous ligaments.

8. The posterior zygapophyseal joints (facets) are well innervated by the articular branches of the posterior primary divisions and by sympa-

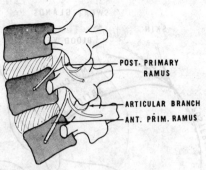

POST. PRIMARY
RAMUS

ARTICULAR BRANCH
ANT. PRIM. RAMUS

Figure 3-21. Innervation of posterior articulations (facets). Lateral oblique views.

thetic nerves and, like any synovial joint, are sources and sites of pain when injured or inflamed (Fig. 3-21).

9. The deep muscles of the spine have been confirmed as sites of pain. In reflex contraction occurring as a "protective spasm," these muscles can enhance pain and can ultimately become the major source of pain (Fig. 3-22).

The tissue sites of the functional unit that when injured, irritated, or inflamed are capable of transmitting nociceptive stimuli are depicted in Figure 3-23. They are summarized as follows:

1. Anterior and posterior longitudinal ligaments
2. Outer layers of the annulus
3. Dura of the nerve roots
4. Capsule of the facets
5. Interspinous ligaments
6. Posterior spinal erector muscles

The basis for injuring, insulting, inflaming, or otherwise adversely affecting the foregoing tissues with resultant pain and impairment is found by the evaluation of faulty body mechanics based on history and physical examination.

Before the faulty body mechanics that initiate nociceptive stimuli are analyzed, *normal* body mechanics must be clarified.

NORMAL BODY MECHANICS

The erect spine during standing can be called posture. The four physiologic curves of the spine have been discussed, but in this context only the five lumbar vertebrae will be considered, as they balance on the sacral base upon the pelvis and both hip joints (Fig. 3-24; see Fig. 3-3).

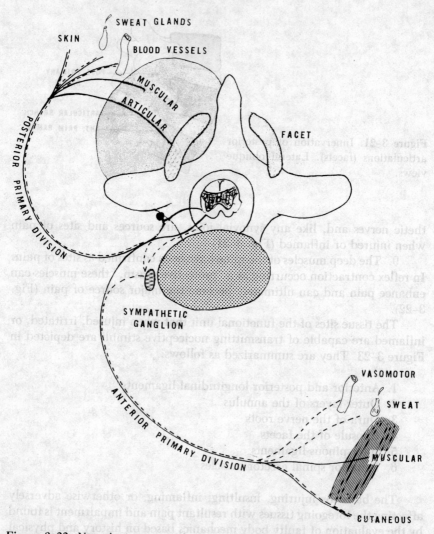

Figure 3-22. Neurologic pain pathway. A simplistic neurologic pathway is depicted, showing anterior and posterior primary rami through which pain sensation is mediated.

To stand erect with all five functional units having flexibility, an effortless neuromusculoskeletal balance must be assumed. A requirement of constant muscular effort would be fatiguing, stressful, and eventually nonproductive. It has been confirmed electromyographically that man stands erect with no muscular effort of the erector spinae other than maintenance of tone. When electromyography (EMG) is used on a person who is standing, no electrical activity can be elicited in the erector spinae muscles other

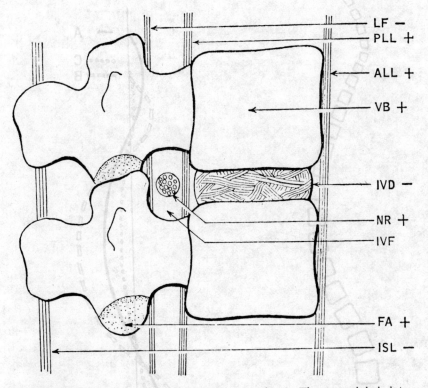

Figure 3-23. Pain-sensitive tissues of the functional unit. The tissues labeled + are pain-sensitive in that they contain sensory nerve endings that are capable of causing pain when irritated. Tissues labeled − are devoid of sensory nerve endings.

  IVF—intervertebral foramen containing nerve root (NR)
   LF—ligamentum flavum
   NR—nerve root
  PLL—posterior longitudinal ligament
  ALL—anterior longitudinal ligament
  IVD—annulus fibrosus of intervertebral disk
   FA—facet articular cartilage
  ISL—interspinous ligament
  VB—ventral body

than occasional activity as the body sways and requires postural reflexes to regain the erect position above the center of gravity.

    The erect posture is thus maintained by the intrinsic pressure within the disks that separate the vertebrae, making the anterior and posterior longitudinal ligaments taut. The erector spinae muscles are slack, as are the posterior longitudinal ligaments between the transverse processes and the posterior superior spine.

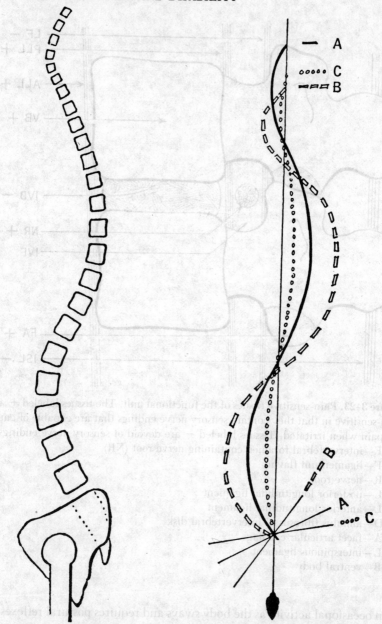

Figure 3-24. Static spine considered erect posture (relationship of physiologic curves to plumb line of gravity). *Left,* Lateral view of the upright spine with its static physiologic curves depicting posture. *Right,* The change in all superincumbent curves as influenced by change in the sacral base angle. All curves must be transected by the plumb line to remain gravity-balanced. Physiologic angle (A), increased angle (B), and decreased sacral angle with flattened lumbar lordosis (C) are depicted.

There is a physiologic lordosis that is dependent on the sacral base. This lumbosacral angle has been depicted in Figure 3–5 and is influenced by the capsular tissues of the hips.

In erect stance, the hips are extended to their physiologic limits by using the iliopectineal ligament. This ligament is essentially a thickening of the anterior portion of the hip joint capsule and termed the **Y ligament of Bigelow.** In the extended hip position, the pelvis rotates posteriorly to "lean" upon this ligament (Fig. 3–25). The lordotic lumbar spine "leans"

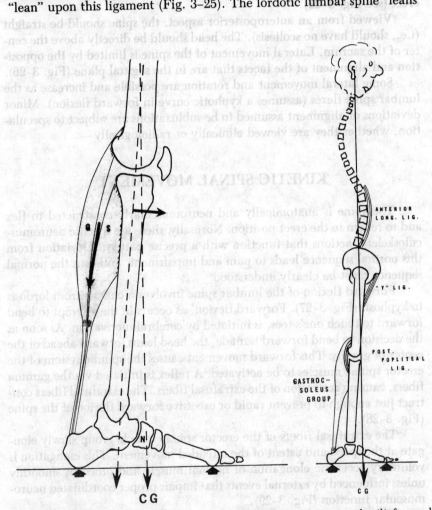

Figure 3–25. Static spine support. The relaxed person leans on the iliofemoral ligament ("Y" ligament of Bigelow), the anterior longitudinal ligament, and the posterior knee ligaments. The ankle cannot be "locked," but when a person leans forward only a few degrees, the gastrocnemius must contract to support the entire body. Relaxed erect posture is principally ligamentous, with only the gastrocnemius-soleus muscle group active. (CG—center of gravity)

upon the anterior longitudinal ligament and to a slight degree upon the posterior zygapophyseal (facet) joints.

The leg extends at the knee and thus "leans" on the posterior popliteal capsule. This extended knee can support the erect body indefinitely using only the ligaments. No muscular activity is required. Only the foot ankle articulation cannot be locked by ligamentous support. At this joint, the gastrocnemius and soleus muscles must maintain tone to keep the lower leg erect (see Fig. 3–25).

Viewed from an anteroposterior aspect, the spine should be straight (i.e., should have no scoliosis). The head should be directly above the center of the sacrum. Lateral movement of the spine is limited by the opposition and alignment of the facets that are in the sagittal plane (Fig. 3–26).

Some lateral movement and rotation are possible and increase as the lumbar spine flexes (assumes a kyphotic curve in forward flexion). Minor deviations of alignment assumed to be subluxations are subject to speculation, whether they are viewed clinically or radiologically.

KINETIC SPINAL MOVEMENT

The spine is anatomically and neuromuscularly constructed to flex and to return to the erect position. Normally, these are specific neuromusculoskeletal actions that function with a precise pattern. Deviation from this normal sequence leads to pain and impairment, and thus the normal sequence must be clearly understood.

Forward flexion of the lumbar spine involves a change from lordosis to kyphosis (Fig. 3–27). Forward flexion, as occurs in the attempt to bend forward to touch one's toes, is initiated by cerebral motivation. As soon as the decision to bend forward is made, the head leans forward ahead of the center of gravity. This forward movement causes the spindle system of the erector spinae muscles to be activated. A reflex is initiated via the gamma fibers, causing activation of the extrafusal fibers. The extrafusal fibers contract just enough to prevent rapid or excessive forward flexion of the spine (Fig. 3–28).

The extrafusal fibers of the erector spinae muscle group slowly elongate at the speed and extent of the intended movement. This elongation is voluntary eccentric elongation of normal muscle and proceeds smoothly unless influenced by external events that impair proper coordinated neuromuscular function (Fig. 3–29).

The extrafusal fibers elongate until they have reached the extent of their fascial sheath flexibility (Fig. 3–30). At this point of elongation, the functional unit has flexed to its maximum. The ligaments between the posterosuperior spines and the transverse process now have also been fully elongated and stop further flexion. The limitation of flexion of each func-

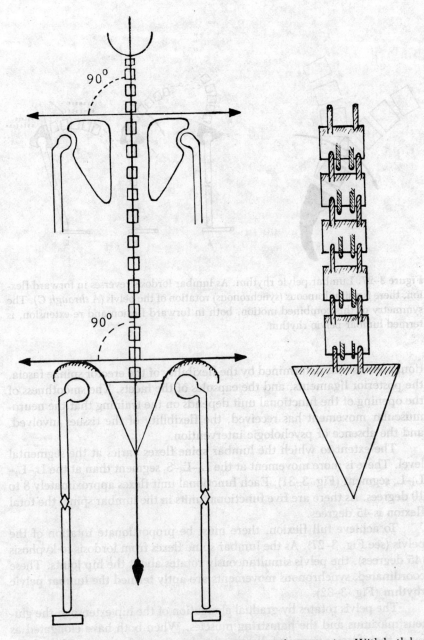

Figure 3–26. *Left*, Anteriorposterior view of the erect human spine. With both legs equal in length, the spine supports the pelvis in a level horizontal plane, and the spine, taking off at a right (90-degree) angle, ascends in a straight line. *Right*, Enlarged drawing shows parallel alignment and proper symmetry of facets in this erect position.

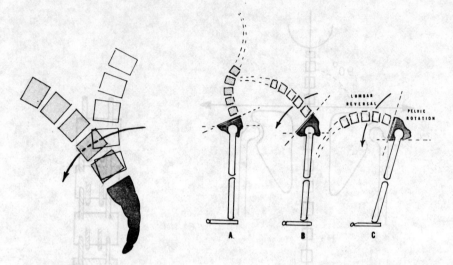

Figure 3-27. Lumbar pelvic rhythm. As lumbar lordosis reverses in forward flexion, there is a simultaneous (synchronous) rotation of the pelvis *(A through C)*. The symmetry of this combined motion, both in forward flexion and re-extension, is termed lumbar pelvic rhythm.

tional unit is thus determined by the flexibility of the erector spinae fascia, the posterior ligaments, and the capsules of the facets. The smoothness of the opening of the functional unit depends on the training that the neuromuscular movement has received, the flexibility of the tissues involved, and the absence of psychologic intervention.

The extent to which the lumbar spine flexes varies at the segmental level. There is more movement at the $L_4-L_5-S_1$ segment than at the $L_1-L_2-L_3-L_4$ segment (Fig. 3-31). Each functional unit flexes approximately 8 to 10 degrees. As there are five functional units in the lumbar spine, the total flexion is 45 degrees.

To achieve full flexion, there must be proportionate rotation of the pelvis (see Fig. 3-27). As the lumbar spine flexes from lordosis to kyphosis (45 degrees), the pelvis simultaneously rotates about the hip joints. These coordinated, synchronous movements are aptly termed the **lumbar pelvic rhythm** (Fig. 3-32).

The pelvis rotates by gradual elongation of the hip extensors: the gluteus maximus and the hamstring muscles. When both have elongated as far as their fasciae permit, full pelvic lumbar flexion has occurred, and this is as far as the person will flex. Full flexion thus depends on the flexibility of the individual, which is determined partly by genetics, but more importantly by the conditioning through exercise that the individual has undergone.

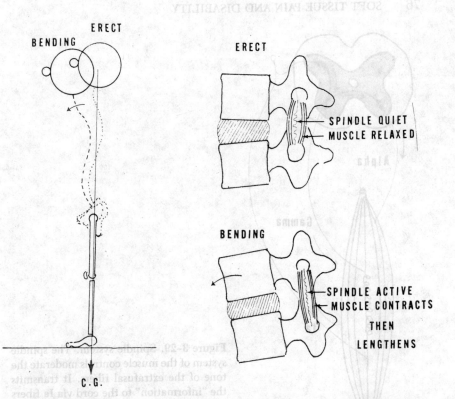

ERECT

BENDING

ERECT

SPINDLE QUIET
MUSCLE RELAXED

BENDING

SPINDLE ACTIVE
MUSCLE CONTRACTS
THEN
LENGTHENS

C.G.

Figure 3-28. Initiation of forward flexion. Once the decision of "bending over" is made, the head goes ahead of the center of gravity (C.G.). The intrafusal fibers of the erector spinae muscles (in each functional unit), which have been dormant in erect posture, are activated by elongation of the spindle system. A reflex mechanism ensues: the spindle system activates the extrafusal fibers (that have also been dormant in the erect stance posture), and these fibers gradually elongate to "open" the functional units, allowing the lumbar spine to flex.

Once the pelvis is fully flexed, as it is in the attempt to touch one's toes, the pelvis has undergone its maximum rotation, and the lumbar spine has flexed to full kyphosis.

To resume the full erect position, the opposite movement must occur—that is, the pelvis must "derotate," and the lumbar spine must simultaneously return to straightening and then resume its lordosis (Fig. 3–33). To do this properly, the pelvis *must* derotate first and almost fully before the lumbar spine resumes its lordosis. In this natural sequence, whether the person is merely resuming the erect posture or lifting an object, the tissue stress on the lumbar spine is upon the fascia of the erector spinae muscles and the posterior ligaments, rather than upon the erector

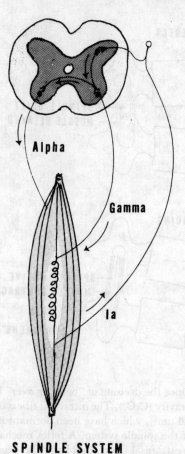

Alpha

Gamma

Ia

SPINDLE SYSTEM

Figure 3–29. Spindle system. The spindle system of the muscle controls moderate the tone of the extrafusal fibers. It transmits the "information" to the cord via Ia fibers at the cord level, where it modifies the tone of the extrafusal muscle fibers (innervated by somatic alpha fibers). The spindle system is "reset" to the appropriate tone via the gamma fibers.

spinae muscles themselves. With this angulation, these connective tissues can sustain this stress, whereas the muscles cannot.

Once having returned to approximately 45 degrees of flexion, the lumbar spine undergoes kyphosis to gradual lordosis. This last change in the lumbar spine is accomplished by shortening of the erector spinae muscles.

Trunk rotation, made possible by flexing the lumbar spine, which thus separates the facet joints, allows considerable range of motion. Return to the erect position requires simultaneous "derotation" of the lumbar spine with re-extension (see Fig. 3–33).

Physiologic flexion, re-extension, rotation, and derotation require flexibility of the tissues and proper neuromuscular control (Fig. 3–34). The control must become "habit," through training and practice, to permit the person to flex and re-extend properly, efficiently, and without pain.

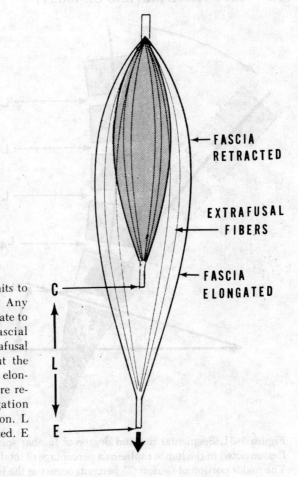

FASCIA
RETRACTED

EXTRAFUSAL
FIBERS

FASCIA
ELONGATED

Figure 3-30. Fascial limits to muscular elongation. Any muscle bundle can elongate to the extent that its fascial sheath permits. The extrafusal fibers elongate fully, but the fascia must be passively elongated. Fascial contracture restricts muscular elongation and joint range of motion. L = length. C = contracted. E = elongated.

STATIC AND KINETIC LOW BACK PAIN

Low back pain results essentially from inflammation, injury, irritation, misuse, or abuse of the tissues within the functional unit of the spine that are capable of producing nociceptive stimuli. These actions decrease the membrane potential of sensory nerve endings, which then transmit sensation to the dorsal horns of the cord for ultimate transmission and translation as pain.

As previously stated, pain can be classified as acute, recurrent, or chronic, and as **static** or **kinetic.** Proper diagnosis of the cause of low back pain requires that the specific meanings of these terms are understood and that the terms are properly used.

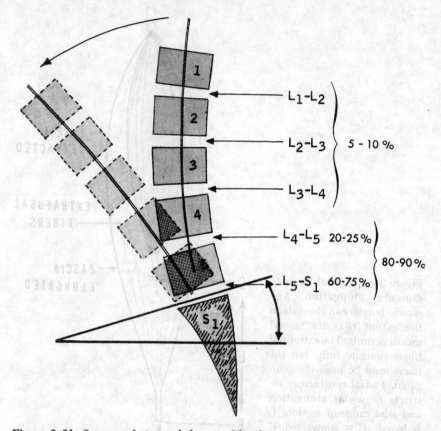

Figure 3–31. Segmental site and degree of lumbar spine flexion. The degree of flexion noted in the lumbar spine as a percentage of total spine flexion is indicated. The major portion of flexion (75 percent) occurs at the lumbosacral joint; 15 to 20 percent of flexion occurs at the L_4–L_5 interspace; and the remaining 5 to 10 percent is distributed between L_1 and L_4. The forward-flexed part of the diagram indicates the mere reversal past lordosis of total flexion of the lumbar curve.

Static Low Back Pain

The erect nonmoving (static) spine can cause low back pain. There are many schools of theory regarding definitions of physiologic and pathologic static low back pain. Normally, the erect spine constitutes posture, and lordosis is physiologic. Excessive lordosis has been generally considered to contribute to low back pain. More recently, excessive kyphosis sustained for long periods has been considered a source of low back pain. Both beliefs merit discussion, as neither alone satisfactorily explains the occurrence of static low back pain.

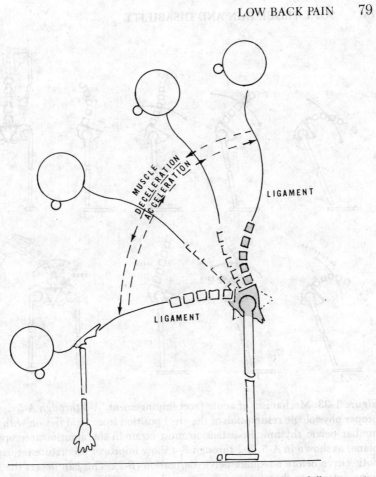

Figure 3-32. Muscular deceleration and acceleration of the forward-flexing spine from the erect ligamentous support to the fully flexed ligamentous restriction. Eccentric and concentric muscular contraction permits forward flexion and re-extension.

Lordotic Low Back Pain. Excessive lordosis, as postulated by Williams,[10] has for years been considered a major cause of low back pain. Reduction of the lordosis has benefitted patients with this type of low back pain.

The functional basis for regarding lordosis as a cause of back pain is that an excessively lordotic posture places more weight on the facets, which are not predominantly weight-bearing joints, but are sites of nociceptive tissue. Also, excessive lordosis narrows the intervertebral foramen by approximating the pedicles. This action compresses the nerve roots and their dural sheaths. The disk in this lordotic posture is compressed poste-

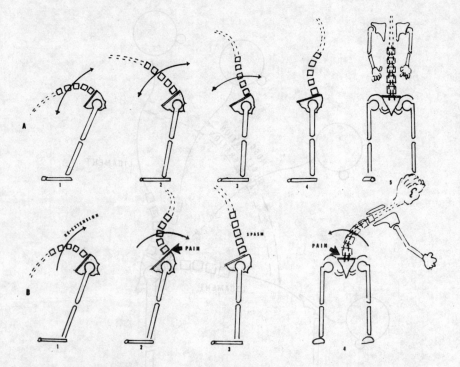

Figure 3–33. Mechanism of acute facet impingement. *A-1 through A-5* depict the proper physiologic resumption of the erect position from total flexion with reverse lumbar pelvic rhythm. Re-extension must occur in the anterior anteroposterior plane, as shown in *A-5*. *B-1 through B-4* show improper premature return of lordotic curve before adequate pelvic derotation *(B-2)*. This cantilevers the lumbar spine anterior to the center of gravity and approximates the facets, causing pain *(B-2)* and spasm *(B-3)*. With the body flexed and rotated, there is further asymmetry of the facets, facilitating unilateral impingement *(B-4)*.

riorly, which theoretically causes the nucleus to migrate forward. There is also some bulging of the posterior annulus in this lordotic posture (as revealed by myelography performed with the patient so extended) (Fig. 3–35). The combination of all these factors imposed upon tissues that are irritated and inflamed from other sources of trauma results in pain.

McKenzie,[11] a New Zealand physical therapist, denied this concept and postulated that man spends much time in the forward flexed position: prolonged sitting while bending forward, or standing while leaning forward. This lumbar kyphotic posture elongates the posterior ligamentous, fascial, and muscular tissues, causing mechanical and thus irritating strain. He postulates that in this position, the intervertebral disks open posteriorly, causing the nucleus to bulge posteriorly. This bulging in turn

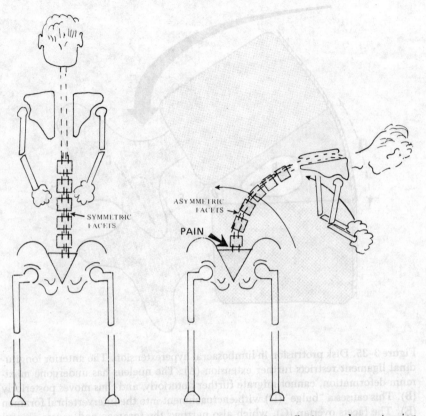

SYMMETRIC FACETS

ASYMMETRIC FACETS

PAIN

Figure 3–34. Asymmetric facets. Pain can occur from flexion and re-extension when the facets are anatomically asymmetric, or when there is faulty derotation.

elongates the posterior annular fibers and encourages disk bulging or posterior herniation (Fig. 3–36), which irritates the sensitive posterior longitudinal ligament and/or the nerve root dura. This theory justifies extension (lumbar lordotic) exercises and posture.

Both theories regarding the cause and treatment of low back pain remain unproven, but clinical experience regarding which posture has been prevalent in the onset of low back pain and which exercises (kyphotic or lordotic) are of benefit should clarify which is desirable. Both theories have validity.

The diagnosis of static low back pain requires that the patient reports pain to occur from a sustained erect posture and that the pain is reproduced during the examination. Sustained excessive lordosis that reproduces the low back pain gives validity to the first concept. Reproducing the low back pain from excessive or sustained forward flexion gives validity to the McKenzie concept.

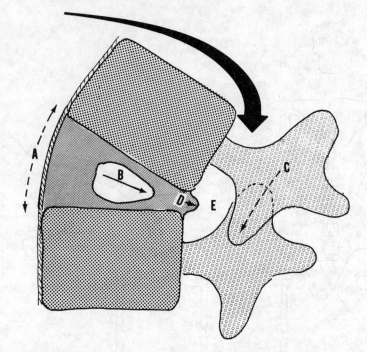

Figure 3–35. Disk protrusion in lumbosacral hyperextension. The anterior longitudinal ligament restricts further extension (A). The nucleus has undergone maximum deformation, cannot migrate further anteriorly, and thus moves posteriorly (B). This causes a "bulge" (D) with encroachment into the intervertebral foramen (E). The facets overlap (C), which also narrows the foramen and causes painful weight bearing.

In severe degenerative disk disease, spondylolisthesis, or facet tropism, extension, which places pressure upon the facets, is diagnostic and indicates precise treatment. Radicular pain (i.e., radiating root pain) can result from excessive lordosis (Fig. 3–37). In this attitude, the nerve root within the foramen is compressed and causes pain.

Kinetic Low Back Pain

By far, the most prevalent cause of low back pain is kinetic. In kinetic low back pain, improper body mechanics are used. This type of pain applies to patients who "injure" the lower back by bending over, returning to the erect stance, lifting, and so forth.

Proper body mechanics have been discussed in detail in a previous section. If a person returns to the erect posture improperly, any of the sensitive tissues of the functional unit can be "insulted." "Improperly"

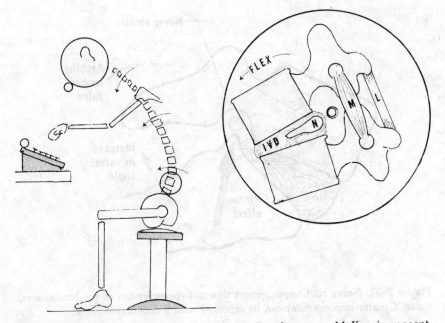

Figure 3-36. Disk "bulging" from prolonged flexed posture: McKenzie concept. McKenzie postulated that man spends much time in the flexed position (e.g., standing at work or sitting), which overstretches the posterior tissues (muscles, ligaments, capsules) but especially pushes the nucleus posteriorly and causes disk bulging. The remedy according to McKenzie is to institute extension (recreate the lordosis).

means that the lordosis is regained prematurely, before the pelvis has completely or synchronously derotated. Thus, the lordosis is regained with the upper body ahead of the center of gravity. All the mechanisms of lordotic pain discussed in the previous section "Static Low Back Pain" apply to kinetic low back pain, but now there is added weight upon the lumbosacral elements due to their being ahead of the center of gravity (Fig. 3–38).

Improper derotation on bending over and returning to the erect posture compounds the problem. In the erect posture, the facets are symmetrically aligned and allow little lateral or rotational movement. As the person bends over, the facets "open," allowing slight rotation and lateral flexion. As the person bends over and twists to one side, the facets open on the convex side of the bending and forward flexion. The facets on the concave side (toward which the person is bending) remain in close approximation.

Excessive or erratic bending over and twisting can damage the tissues on the convex side by excessive elongation. Upon re-extending, more trouble occurs. Unless the person smoothly and synchronously "derotates" dur-

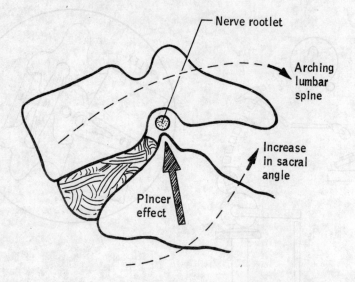

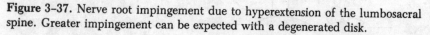

Figure 3–37. Nerve root impingement due to hyperextension of the lumbosacral spine. Greater impingement can be expected with a degenerated disk.

Figure 3–38. Proper and improper bending and lifting. Regaining the lower back lordosis too soon in lifting, and with the knees not bent, causes low back pain.

ing re-extension to the erect posture, the facets on the concave side can "lock," causing functional unit rotation at this site. Rotation about the new axis imposes shear on the anterior disk and excessive motion on the facet of the convex side. Either factor can cause tissue damage with resultant pain.

This improper re-extension from the forward bent and rotated position causes low back pain from the following possible factors:

1. The facets on the concave side "lock," with resultant synovitis, a capsular entrapment, or a mild subluxation of the facet joint.
2. The facet joint on the convex side of the flexed position exceeds capsular laxity, and this joint can sublux, causing acute synovitis from capsular stretching or tearing.
3. The intervertebral disk undergoes excessive shear and rotation as the axis of functional unit rotation is now at the concave facet. The annular fibers can tear, as can the longitudinal ligaments, causing low back pain. Some bulging of the annulus can result, causing pressure on the nerve root of that foramen and causing pain to radiate to the leg (Fig. 3–39, 3–40).

Diagnosis of kinetic low back pain is derived essentially from the history. How the patient bent over and returned to the erect stance must be clearly delineated. Did the patient bend over and twist to one side? Was the object reached for far from the midline? Was the lifted object heavy or bulky? Was the person thinking of what he or she was doing? Was the person distracted? Was the person tired, angry, despondent, depressed, anxious, or impatient?

All these questions relate to whether proper body mechanics were employed during bending over, re-extending, and lifting. If the person was properly trained and used proper body mechanics, why did faulty lifting or re-extension occur? No benefit can be derived from conditioning of the lower back tissues, good flexibility, strength, and proper alignment if the person uses improper body mechanics. This point will be further discussed in the section on treatment.

Upon re-extension or improper lifting of the impinged tissues, whether they are ligaments, joint capsules, or outer annular fibers, nociceptive stimuli are liberated at the injured tissue site. Immediate protective contracture of muscle—"spasm"—occurs. This spasm results from a combination of a neurologic reflex and a chemical release within the tissues that irritates the local muscles.

Clinically, the patient complains of localized pain. The erect spine assumes an "antalgic" posture, that is, a flat low back with no lordosis. The patient is unable to bend over or sit, owing to the fact that the functional units of the low back cannot flex: the erector spinae muscles will not

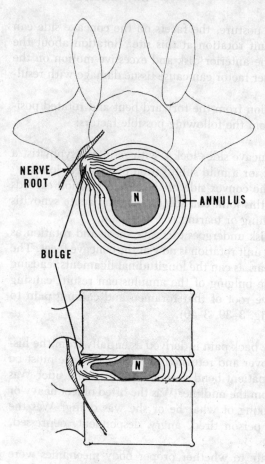

Figure 3–39. Disk (nucleus) bulging. Some degree of disk bulging occurs daily, physiologically. It becomes pathologic and pain-producing when there is some annular deficit that permits the nucleus to bulge further than its normal degree. A bulging disk viewed from above is depicted (*top*). There are still outer annular fibers; thus, the disk has not "ruptured."

The same disk bulging viewed from the side is depicted (*bottom*). Both views show approachment of the bulge toward the nerve root.

If the tissue injury has been unilateral, the muscles of that side undergo spasm, and the patient assumes a functional scoliosis (Fig. 3–41).

In the examination, a straight leg raising (SLR) test is indicated and must be properly interpreted. A "muscular positive" result of the SLR test occurs when the hamstrings pull upon the pelvis, and in attempting to rotate the pelvis, cause low back pain. This pain results when both legs are raised. A "neurologic" result of the SLR test occurs, as a rule, when one leg is raised and there are positive dural signs. As the leg is raised to the point of pain, the pain can be accentuated by flexing the neck or dorsiflexing the foot at the ankle. These two maneuvers stretch the dura of the nerve root within the foramen and indicate nerve pain.

The SLR test, commonly called the Lasegue test, yields positive or negative results depending on whether the nerve root dura is inflamed. The test was originated by Forst[4] and later described by Lasegue.[5] Both assumed the test to indicate hamstring muscle spasm. Later, it was found

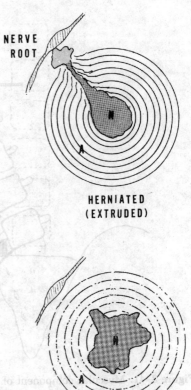

NERVE
ROOT

HERNIATED
(EXTRUDED)

DEGENERATED

Figure 3–40. Disk herniation and degeneration. *Top,* An extruded nucleus in which the nucleus has "herniated" out of the torn annular fibers. *Bottom,* Degenerated nucleus with fragmented annular fibers.

that the nerve moved within the intervertebral foramen as the leg was flexed.

In patients who have no inflammation of the nerve, no pain is evoked by mere touch. The work of Lindahl,[6] Granit and associates,[7] and McNab[8] clarifies these findings. An inflamed dura is necessary to produce a "positive" SLR result (i.e., is a "positive dural sign") (Fig. 3–42).

A neurologic examination must always be performed. Each of the dermatomes and the myotomes of the lumbosacral plexus can and must be tested (Fig. 3–43).

The S_1 root innervates the ankle jerk reflex. The strength or weakness of this reflex is difficult to quantitate. It is usually either absent or present and must be compared to the opposite ("normal") leg. The S_1 root is best tested by testing the strength and endurance of the gastrocnemius and soleus muscles. This is achieved simply by having the person *rise up and down numerous times on the toes.* Fatigue is more revealing than mere

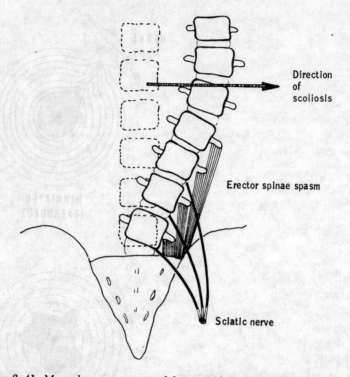

Figure 3–41. Muscular component of functional scoliosis. The deep interspinous muscles, including the multifid muscle, act unilaterally to pull the spine and create a functional scoliosis. Spasm of the muscle, considered to splint and thus be protective, is thought to be involved.

testing of one or two elevations or of walking across the room on the tiptoes (Table 3–1).

The S_1 root can also be tested by testing the gluteus maximus and gluteus medius. Spreading the legs apart against resistance tests the gluteus medius. With the hip and knee flexed, and the foot upon the ground or examining table, elevating the buttocks from the floor several times tests the endurance of the gluteus maximus.

The L_5 myotome is best tested by resisting the elevation of the big toe. Again, repeated "picking up" of the big toe (extensor hallucis longus) tests the endurance and integrity of the L_5 myotome.

The myotomes and the dermotomes are each contained in a specific nerve root as depicted in Figure 3–44.

The anterior tibialis is tested by resisting ankle dorsiflexion with the foot inverted; again, this action is performed repeatedly. This procedure tests the L_5 and a portion of the L_4 root myotomes.

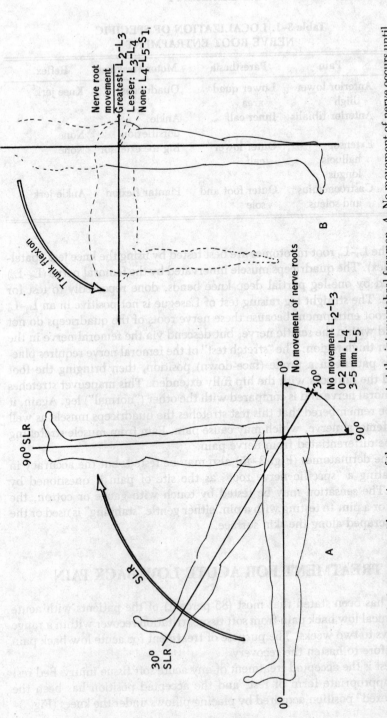

Figure 3–42. Straight leg raising with motion of the nerve root at the intervertebral foramen. *A,* No movement of nerve occurs until 30 degrees of straight leg raising, following which there is movement of up to 2 mm at L_4 root, and 3 to 5 mm at L_5 root. *B,* Movement of nerve roots and foramina on forward trunk flexion is depicted.

Table 3–1. LOCALIZATION OF SPECIFIC
NERVE ROOT ENTRAPMENT

Root	Pain	Paresthesia	Motor Weakness	Reflex
L_3	Anterior lower thigh	Lower quad area	Quadriceps	Knee jerk
L_4	Anterior tibialis	Inner calf	Ankle dorsiflexion	None
L_5	Extensor hallucis longus	Outer lower calf	Big toe extensor	None
S_1	Gastrocnemius and soleus	Outer foot and sole	Plantar flexion	Ankle jerk

The L_3–L_4 root myotomes are best tested by using the knee jerk (patellar reflex). The quadriceps muscle innervated by the femoral nerve (L_3–L_4) is tested by one-leg partial deep knee bends, done repeatedly to test for fatigue. The straight leg raising test of Lasegue is not positive in an L_3–L_4 nerve root entrapment because these nerve roots of the quadriceps do not descend within the sciatic nerve, but descend via the femoral nerve in the anterior thigh region. The "stretch test" of the femoral nerve requires placing the patient in a prone (face-down) position, then bringing the foot toward the buttocks with the hip fully extended. This maneuver stretches the femoral nerve and is compared with the other ("normal") leg. Again, it must be remembered that this test stretches the quadriceps muscle as well as the femoral nerve, which may cause pain. Pain from muscle stretching must be differentiated from nerve pain.

The dermatomes (Fig. 3–45) also may be tested, but the accuracy in designating a "specific nerve root" as the site of pain is questioned by many. The sensation may be tested by touch with gauze or cotton, the finger, or a pin. In testing with a pin, either gentle "stabbing" is used or the pin is scraped along the skin surface.

TREATMENT FOR ACUTE LOW BACK PAIN

It has been stated that most (85 percent) of the patients with acute mechanical low back pain from soft tissue irritation recover within a range of 5 days to two weeks. The purpose of treatment for acute low back pain is therefore to hasten the recovery.

Rest is the accepted treatment of any acute soft tissue injury. Bed rest is the appropriate form of rest, and the accepted position has been the "semiflexed" position acquired by placing pillows under the knees (Fig. 3–

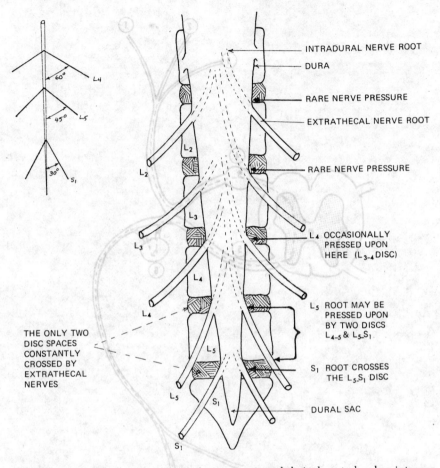

Figure 3-43. Relationship of spinal nerve roots and their dura to lumbar intervertebral disks. Site of nerve root compression is indicated.

46). The average head pillow completes the flexed position. It must be stated, however, that *the best position is the position of comfort.*

Oral medications also aid in relieving the acute pain. Assuming that nociceptive tissue chemicals are initiated at the site of injury (see Chapter 2), the anti-inflammatory drugs, if tolerated and not contraindicated, may play a favorable role. Salicylates are the time-tested anti-inflammatory drugs; the current nonsteroidal drugs may be used when acetylsalicylic acid (ASA) is not effective. Muscle relaxants and tranquilizers are useful when there is a great deal of anxiety and intolerance of pain associated with the injury. Reassurance and avoidance of "labels that are ominous" are better methods of reducing the patient's anxiety.

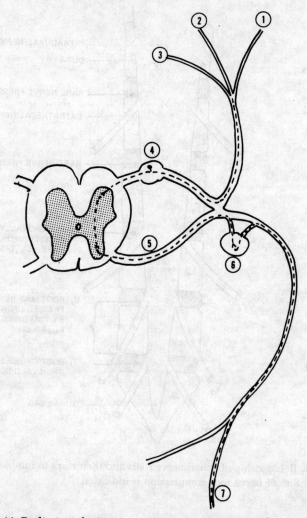

Figure 3-44. Radiation of sciatic nerve into anterior primary division with noxious irritation of posterior skin region (1); erector spinae musculature (2); posterior articulations (facets) (3). Radiation proceeds down the sensory root (4) and down the anterior primary division (5), gathering sympathetic innervation (6), with dermatomal distribution down anterior primary division (7).

The duration of necessary "rest" is being questioned. Immobilization for too long a period causes muscle atrophy and soft tissue contracture as well as anxiety, depression, and dependence. Gentle bed exercises done isometrically at first are valuable. Gradual addition of isotonic exercise can be beneficial.

The relief derived from the local application of ice and heat has a

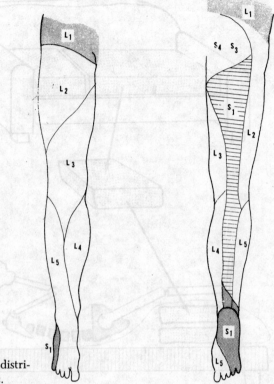

Figure 3-45. Dermatome distribution of lower extremities.

physiologic basis. Ice is analgesic, decreases the vasomotor and inflammatory chemical stimuli at the tissue site of injury, and decreases local muscular spasm. Ice can easily be applied by brushing with an ice applicant (Fig. 3-47). Heat later benefits the injury, as it brings in a fresh supply of blood to "wash out" and replace the toxic metabolites. Also, heat is a sedative and relaxes muscle spasm.

Manipulation has been advocated by many as an effective treatment. This treatment is based on the assumption that "something is out of alignment." Although malalignment has not yet been verified, manipulation proves effective in many cases, *provided that a meaningful examination precedes the treatment* and rules out neurologic impairment or pathologic tissues. Manipulation usually is effective for a few sessions, but repeated manipulations that fail to afford relief should be questioned, and the patient should be re-evaluated.

Anesthetic and steroidal injections have been beneficially used. The nociceptive stimuli of the injured tissue can be minimized by injection of an anti-inflammatory agent into the site of injury. The resultant muscle spasm can be diminished by an appropriate injection of an anesthetic

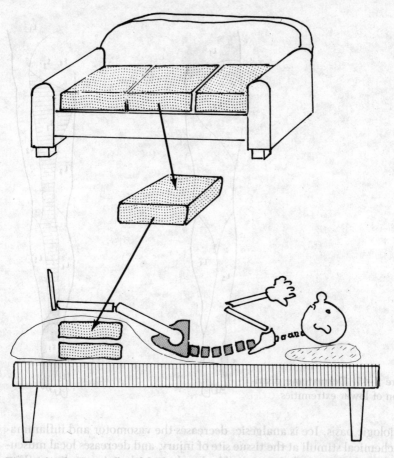

Figure 3–46. Home flexed-bed posture. Square sofa pillows are inserted under bedsheets to facilitate flexed hip and knee posture in lumbar disk disease.

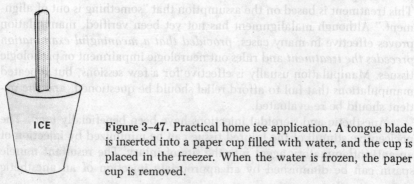

Figure 3–47. Practical home ice application. A tongue blade is inserted into a paper cup filled with water, and the cup is placed in the freezer. When the water is frozen, the paper cup is removed.

agent. As in manipulation, benefit is derived from a few injections and rarely is "specific": it is merely palliative.

After a period of bed rest in "the most comfortable position," usually the flexed position, followed by application of ice, then heat, the patient must then begin resuming the erect posture.

Usually, bed rest ranges from 3 to 4 days or from one week to 10 days. If pain persists longer than 10 days, a re-evaluation of the patient is indicated. Severe pain and persistent scoliosis, abnormal straight leg raising, and/or neurologic deficit indicates the possibility of sciatic neuropathy, which means that further bed rest, stronger medications, and conceivably other modes of treatment are needed.

When symptoms improve, simple bed exercises should be slowly instituted. Gently bringing one knee to the chest, then the other knee, and then both knees slowly stretches the lower back, which has usually been in "spasm." This exercise must be done *slowly* and gently (Fig. 3–48). The knees should not necessarily be brought to the chest the first few times the exercise is attempted.

Exercises that require "bouncing" or forceful stretching are to be condemned, as they usually attempt to stretch muscles that have been contracted and may in turn be inflamed.

Once the lower back appears to be flexible, gentle pelvic tilting exercises can be started. The lower back is pressed against the bed, and the pelvis is "lifted" gently and slowly, held there, and then slowly lowered. The sequence is repeated four to six times. This exercise stretches the lower back and causes the back muscles to "relax reciprocally" as the abdominal muscles contract.

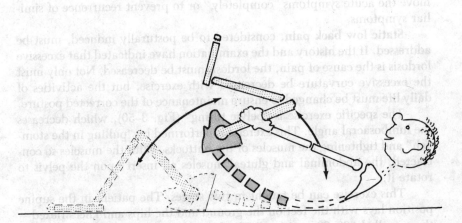

Figure 3–48. Gentle lower back stretching exercises in which the femurs are used for leverage.

Once improvement is made, the patient must assume the erect position. From the supine position, the patient should use the arms to roll to one side and slowly come to a sitting position. The hips and legs remain flexed. Once seated erect, the patient then must place the feet directly under the body and slowly assume the erect posture.

If the lower back continues to be symptomatic, that is, continues to demonstrate pain and limited flexibility, limited upright daily activities continue as further rehabilitation modalities for the lower back are employed.

In some cases, the back remains painful as a result of any duration of being "up" and bending over. A corset may have value in placing the lower back in a comfortable position and in preventing excessive movement (Fig. 3–49). When a lumbosacral back support is used, the principles of bracing must be observed: the upper and lower points of contact must be the thoracolumbar junction and the sacrum, respectively. The abdomen must be supported and "uplifted." A corset must always be used properly—that is, it must be used for a limited period to accomplish the immediate purpose, and its use must be accompanied by proper exercises to allow gradual removal of the corset without recurrence of symptoms. The corset must never become a "way of life" and an object of dependency.

TREATMENT FOR LOW BACK PAIN OF "SPECIFIC" MECHANICAL ORIGIN

Upon cessation of acute symptoms, or even of persistence of some low back symptoms, the precise causative factors must be determined to remove the acute symptoms "completely," or to prevent recurrence of similiar symptoms.

Static low back pain, considered to be posturally induced, must be addressed. If the history and the examination have indicated that excessive lordosis is the cause of pain, the lordosis must be decreased. Not only must the excessive curvature be decreased with exercise, but the activities of daily life must be changed to ensure maintenance of the corrected posture.

The specific exercise is "pelvic tilting" (Fig. 3–50), which decreases the lumbosacral angle. This exercise is performed by "pulling in the stomach" and tightening the muscles of the buttocks. With the muscles so contracted, the abdominal and gluteal muscles all insert upon the pelvis to rotate it.

This exercise can be taught in two stages. The patient in the supine position lays with the feet on the ground and the hips and knees flexed—the so-called "90–90 position." The lower back is then "pressed" to the floor and held there. Placing the patient's hand at the small of the back assures that the lower back has been pressed to the ground. While holding

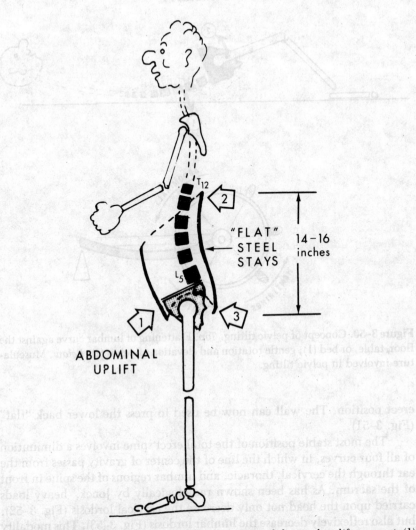

Figure 3–49. Proper lumbar corseting with firm abdominal uplift support (1). Back points of contact are at thoracolumbar junction (2), and over the sacrum (3). Flattened stays decrease lordosis and restrict activities requiring flexion and extension.

this position, the patient slowly raises the buttocks a few inches from the ground. This second stage of the exercise can be done only by contracting the buttocks and the abdominal muscles.

The patient can appreciate the sensation of this pelvis tilt and can continue to use the exercise. As it is mandatory that this "flat-back" posture continues in the erect posture, the exercise should also be done in the

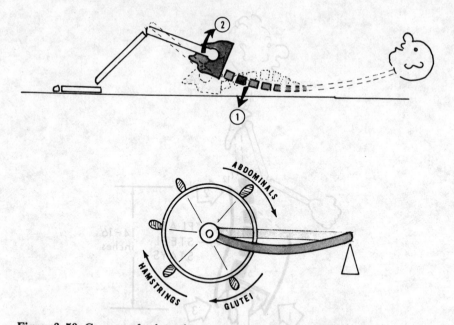

Figure 3–50. Concept of pelvic tilting. *Top,* Flattening of lumbar curve against the floor, table, or bed (1); gentle rotation and elevation of pelvis (2). *Below,* Musculature involved in pelvic tilting.

erect position. The wall can now be used to press the lower back "flat" (Fig. 3–51).

The most stable position of the total erect spine involves a diminution of all four curves, in which the line of the center of gravity passes from the ear through the cervical, thoracic, and lumbar regions of the spine in front of the sacrum. As has been shown radiologically by Jonck,[9] heavy loads carried upon the head not only decrease the cervical lordosis (Fig. 3–52), but also reflexively decrease the lumbar lordosis (Fig. 3–53). This modality of weight upon the head can be used therapeutically to instruct the patient on proper posture. Carrying such a weight decreases a cervical lordosis but also decreases the lumbar lordosis and implants the sensation of proper posture after the weight has been removed in sitting, standing, and walking.

Modifying the daily activities requiring prolonged standing by using proper posture may require the use of a small footstool for one foot or changes in the heights of tables, chairs, and so forth. Modifying the environment may be as important as modifying the person's posture.

If it has been ascertained that lordosis is desirable, and that prolonged or excessive forward bending has been the cause of pain (McKenzie concept), exercises that initially passively "arch" the lower back and gradually

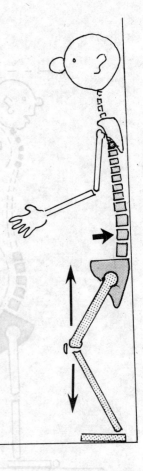

Figure 3-51. Erect pelvic tilting. With the patient standing against a wall with feet slightly forward, the pelvis is "flattened" against the wall, as done against the floor in the supine exercise. This exercise trains the patient to feel the position of the "flat" lumbar spine.

strengthen the extensor muscles must be instituted. Instruction regarding posture and daily activities ensuring slight lordosis must also be given.

Kinetic low back pain is caused by faulty body mechanics. Its treatment involves the following:

1. Regaining full flexibility of the lower back, hamstrings, hip flexors, and gastrocnemius and soleus muscles.
2. Gaining or regaining significant strength of the muscles of the abdomen, buttocks, and hamstrings, and even the extensor back muscles.
3. Learning proper body mechanics in such movements as bending, lifting, and untwisting. Daily practice instills "habit."
4. Evaluating and correcting any psychologic impediment to proper function, such as fatigue, anger, anxiety, impatience, or depression.

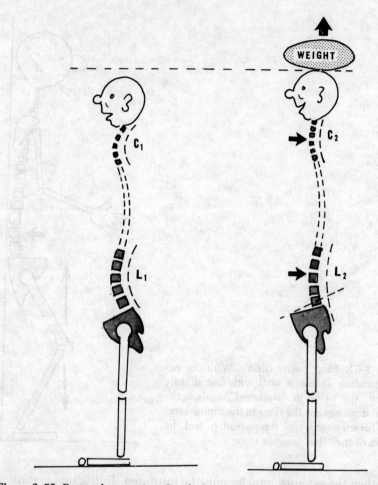

Figure 3-52. Postural correction. By placing a weight upon the head and elevating it, the posture is improved in that the lumbar and cervical curves are decreased and the body approaches the center of gravity.

Should the low back symptoms persist, other modalities, including traction, must be considered. One must always remember that any "other" modality does not preclude or replace the fundamental exercises.

For centuries, traction has been advocated as a treatment of spinal complaints. The basis for using traction in low back pain, and more recently for the "herniated lumbar disk," is hypothetical. Traction benefits patients probably by elongating the spine, decreasing the lordosis, stretching the longitudinal ligament, which allegedly forces any protruding disk back between the adjacent vertebrae, and "opening" the intervertebral foramen by flexing the spine. Also, the facets are separated by the distrac-

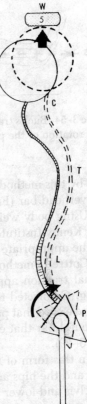

Figure 3-53. Postural change from distraction. The weight placed upon the head causes the pelvis to rotate, which then decreases the lumbar lordosis. The superincumbent curves decrease, and posture is improved.
W—weight
C—cervical vertebrae
T—thoracic vertebrae
L—lumbar vertebrae
P—pelvis

tion forces. The posterior spinous muscles that are in "spasm" allegedly relax, which interrupts the pain cycle.

The original "standard" method of applying traction was Buck's traction: traction forces were applied to the legs, and these forces indirectly elongated the spine (Fig. 3-54). This type of traction was tedious and difficult to maintain, and it did not essentially decrease the lordosis. Much of the traction force was dissipated by friction upon the mattress, and often, by accentuation of the lordosis.

Pelvic band traction (Fig. 3-55) has also proved inadequate, as it loses traction force from friction of the mattress and accentuation of the lordosis. Elevating the entire body via an overhead bar with leg suspension (Fig. 3-56) effectively decreases friction and the lordosis, but applies very little traction forces.

Single-strap pelvic traction is effective and allows larger weights to be used in applying the traction. Forty to 60 pounds of traction can be ap-

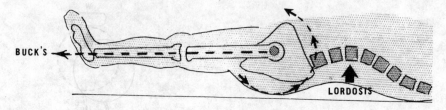

Figure 3–54. Buck's traction. Because the line of pull is anterior to the lumbosacral joint, rotation of the pelvis can increase the lordosis.

plied by this method, and the angle of traction can be varied by elevating the overhead bar (Fig. 3–57).

Using body weight for traction has been employed favorably at the Sister Kenny Institute (Fig. 3–58), but this method requires hospitalization and the appropriate circular bed.

Cottrell's method of traction, the so-called 90/90 position, which uses manual traction applied by the patient, can be beneficial in the home setting. Connected to the pelvic band is a posterior central strap, which leads to a cord that passes through a pulley. This pulley is suspended from a horizontal bar that elevates the pelvis when the patient pulls on the cord (Fig. 3–59).

In this form of traction, the legs are placed on a bolster so that the knees and the hips are 90 degrees from the floor. This position also flexes the pelvis and lower back. The head and upper back are placed on a bolster to assure comfort. The extent of traction is varied by the patient. Usually, traction is applied 15 minutes at a time four to five times a day and is increased to as much as 4 hours a day if tolerated by and beneficial to the

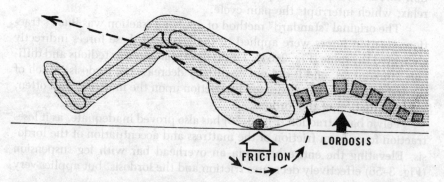

Figure 3–55. Pelvic band traction: friction factor. The pelvic band traction pulls on the pelvis. The friction of the body upon the bed or table causes a pivot point about which the pelvis rotates. This specific type of rotation increases lordosis.

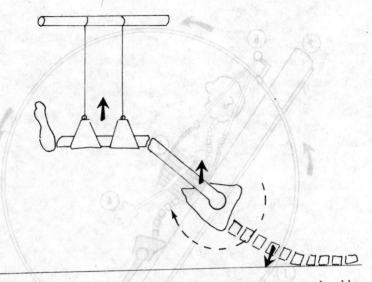

Figure 3-56. Elevation pelvic traction. Suspending the legs from an overhead bar elevates the pelvis, and this decreases lumbar lordosis.

patient. The rope may need to be tied once the amount of traction is determined, as holding it by hand would be inconvenient.

Recently, various techniques of gravity traction have been advocated (Fig. 3-60). Using boots hung over a bar and hanging upside down by the knees are two of these techniques. The head-down position may have side effects of increasing blood pressure and increasing intraocular pressure as well as causing headache, visual blurring, and dryness of the eyes in contact lens wearers. No significant organic damage has been documented

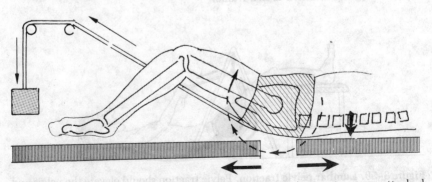

Figure 3-57. Single-strap pelvic traction. With a central posterior strap attached to the pelvic band, the traction pull rotates the pelvis, minimizes friction, and decreases lordosis. A split table, when elongated, further decreases friction.

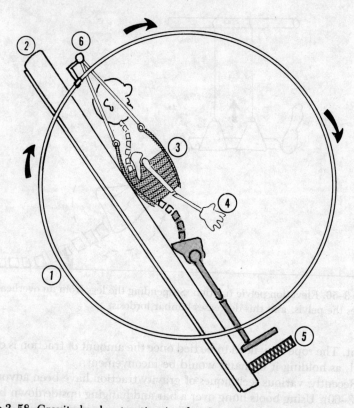

Figure 3–58. Gravity lumbar traction (used at Sister Kenny Institute). 1, Lumbar traction applied in a hospital setting with a circular bed. 2, Mattress with bedboard within Stryker circ-o-lectric bed. 3, Chest harness with lower straps under the rib cage and upper straps firmly grasping the rib cage. 4, Bed manually controlled by patient. 5, Footplate placed several inches below patient's feet for security. 6, Snap ring attached to bed frame.

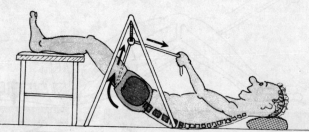

Figure 3–59. Lumbar pelvic traction. Pelvic traction should elevate the pelvis and reverse the lumbar lordosis. There are many forms of traction, but the one depicted has the pelvic band elevating the pelvis with manual assistance. The legs are elevated in a comfortable flexed-knee position. Gravity is thus added to the manual elevation of the pelvis.

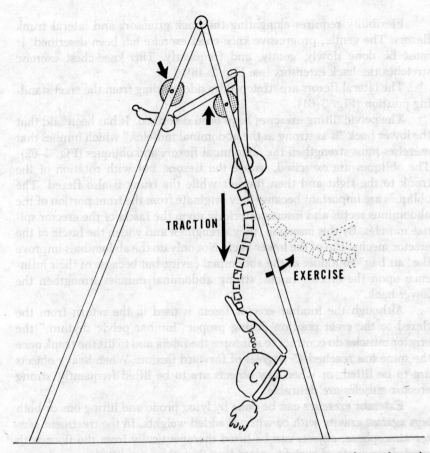

Figure 3-60. Gravity traction stretch exercises. When the patient hangs from the lower extremities, the upper body weight applies traction to the spine. Lateral flexion can be achieved in this manner. Flexion exercises *(dotted lines)* strengthen the abdominal muscles.

from this upside down position, but all of these side effects should be discussed with the patient and evaluated if they occur. Insofar as gravity traction may have limited value, no side effect should be tolerated merely to ensure that the patient undergoes this type of traction.

EXERCISES FOR THE LOWER BACK

Undoubtedly, in most concepts of treatment of lower back pain, exercise is considered the most important factor. The purpose of exercise is to increase flexibility and to improve strength of the pertinent functional muscles.

Flexibility requires elongating the back extensors and lateral trunk flexors. The gentle, progressive knee-chest exercise has been described. It must be done slowly, gently, and frequently. This knee-chest exercise stretches the back extensors (see Fig. 3–48).

The lateral flexors are stretched by side bending from the erect standing position (Fig. 3–61).

The pelvic tilting exercises have been described. It has been said that the lower back "is as strong as the abdominal muscles," which implies that exercises must strengthen the abdominal flexors and obliques (Fig. 3–62). The obliques are exercised, as are the flexors, but with rotation of the trunk to the right and then the left while the trunk is also flexed. The obliques are important because they originate from the front portion of the abdominus rectus and insert posteriorly upon the fascia of the erector spinae muscles. By this insertion, they strengthen and widen the fascia of the erector mechanism of the lower back. Not only do the abdominals improve the "air bag" container of the abdominal cavity, but because of their influence upon the erector fascia, strong abdominal muscles strengthen the lower back.

Although the lumbar erector fascia is used in the return from the flexed to the erect position during proper "lumbar pelvic rhythm," the erector muscles do contract to reinforce the fascia and to lift the trunk once the spine has reached 45 degrees of forward flexion. When heavy objects are to be lifted, or when any objects are to be lifted frequently, strong erector muscles are desirable.

Extensor exercises can be done by lying prone and lifting one or both legs against gravity with or without added weights. In the treatment center atmosphere, objects can be lifted therapeutically from the floor with graded amounts of weight, providing that *they are lifted properly and with careful concentration and effort.*

As objects should be lifted with knees bent, strong quadriceps muscles are desirable. The best exercise for strengthening the quadriceps is a partial deep knee bend done with the pelvis tilted.

Protective hamstring stretches and heel cord stretches should be done frequently to ensure their flexibility. "Tight" heel cords and hamstrings place a strain on the lower back, as they (especially the hamstrings) do not "give," which places a greater burden on the flexing spine.

Other exercises are valuable in giving total conditioning to the body. Arm exercises strengthen the shoulders and the latissimus dorsi muscles, which "protect" and reinforce the lower back.

Exercises generally improve the psychologic "tone" of a person, so that a well-conditioned lower back becomes part of a well-balanced, relaxed, self-controlled person. *No matter how well conditioned the tissues of a person's lower back are, they will not function properly if the person is not psychologically well controlled.* Well-conditioned muscles and ligaments can be misused and abused and result in pain and disability.

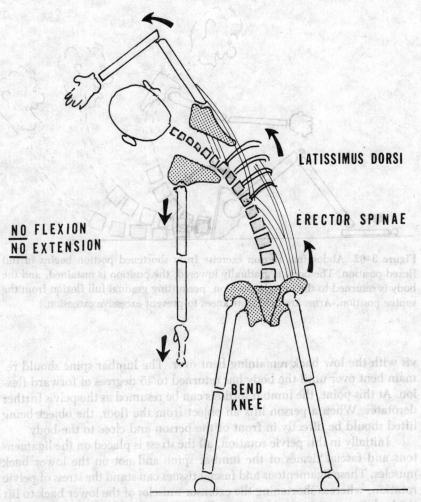

Figure 3–61. Lateral flexion exercises. With the legs apart slightly and one knee bent, the trunk is gently and progressively flexed laterally *without forward or backward flexion.* The unilateral erector spinae muscles and their fascia are stretched. As the scapular muscles also attach to stabilize the spine, they too must be stretched by placing the arms overhead.

REGAINING PROPER BODY MECHANICS

Re-educating the patient in proper body mechanics implies instituting lumbar pelvic rhythm, which has been discussed in an earlier section of this chapter. Neuromuscular re-education is the basis for the current approach to treatment of low back pain.

In returning from a position of bending over, the knees should be bent slightly. The initial "lift" should be accomplished by derotation of the pel-

Figure 3-62. Abdominal flexion exercise from shortened postion begins in full flexed position. The body is gradually lowered, the position is sustained, and the body is returned to the upright position, permitting gradual full flexion from the supine position. Arms are held near knees to prevent excessive extension.

vis with the low back remaining bent over. The lumbar spine should remain bent over until the body has returned to 45 degrees of forward flexion. At this point, the lumbar lordosis can be resumed as the pelvis further derotates. When a person lifts an object from the floor, the object being lifted should be directly in front of the person and close to the body.

Initially in this pelvic rotation, all the stress is placed on the ligamentous and fascial tissues of the lumbar spine and not on the lower back muscles. These ligamentous and fascial tissues can stand the stress of pelvic rotation, whereas shortening the extensor muscles of the lower back to lift the body or an object overwhelms these muscles (Fig. 3-63).

From 45 degrees of flexion, full extension to the erect position combines further pelvic derotation with lumbar lordosis. Through this range, the last 45 degrees, the back extensor muscles are physically capable of bringing the body to an erect position.

It is important that (1) the lumbar pelvic movement is done precisely and smoothly, that is, with voluntary yet automatic effort, which requires training and practice until a good habit results, and (2) the tissues of the low back, the ligaments, the fascia, and the muscles are in good condition from appropriate exercises.

If the person is resuming the erect posture from a bent over and rotated position—that is, if the person is bent over and turns toward one side of the body to lift an object (the object is not directly in front of the per-

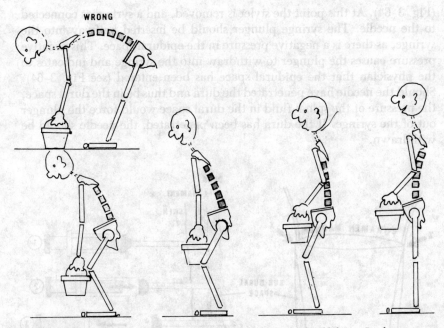

WRONG

Figure 3–63. Incorrect *(top)* and correct *(bottom)* methods of lifting. In the correct method, the spine bends with simultaneous bending of the knees, and the object is held close to the body. The entire body flexes as the object is picked up. The spine re-extends to the upright position with simultaneous extension of hips and knees.

son), proper re-extension and simultaneous derotation must be used. In other words, the low back must re-extend and derotate synchronously.

Often, kinetic low back pain occurs because (1) the lumbar lordosis is resumed prematurely while the body is still in forward flexion ahead of the center of gravity, and (2) the person has re-extended before fully derotating, that is, has regained part or all of the lordosis while still "twisted" to one side. The reason for this improper re-extension has been discussed.

If the low back pain persists because the initial trauma has been excessive, causing severe inflammation of or injury to muscles, ligaments, or fascia, the use of steroids may be beneficial.

Oral steroids such as prednisone or triamcinolone may be administered in large doses for 5 days. If oral steroids cannot be tolerated, epidural steroids may be helpful, especially if there is a radicular component to the pain.

To perform an epidural injection, the patient is placed in a flexed position, and a 17-gauge spinal needle with a Huber point is centrally inserted between the third and fourth vertebral bodies. After the skin has been penetrated, resistance from the interspinous ligament is encountered

(Fig. 3–64). At this point the stylet is removed, and a syringe is connected to the needle. The syringe plunger should be inserted halfway into the syringe, as there is a negative pressure in the epidural space. This negative pressure causes the plunger to withdraw into the syringe and indicates to the physician that the epidural space has been entered (see Fig. 3–64). Should the needle have penetrated the dura and thus be in the dural space, the pressure of the spinal fluid in the dural space would force the plunger out of the syringe. If the dura has been penetrated, the needle should be withdrawn.

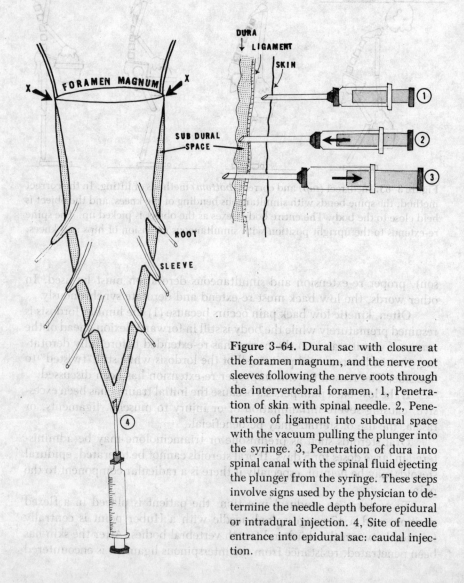

Figure 3–64. Dural sac with closure at the foramen magnum, and the nerve root sleeves following the nerve roots through the intervertebral foramen. 1, Penetration of skin with spinal needle. 2, Penetration of ligament into subdural space with the vacuum pulling the plunger into the syringe. 3, Penetration of dura into spinal canal with the spinal fluid ejecting the plunger from the syringe. These steps involve signs used by the physician to determine the needle depth before epidural or intradural injection. 4, Site of needle entrance into epidural sac: caudal injection.

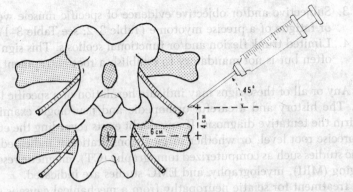

Figure 3-65. Technique for nerve block. Injection is placed lateral to the midpoint between two posterior spines at a distance approximately 6 cm from midline. The needle is directed toward the midline at a 45-degree angle.

Upon ascertaining that the needle is in the epidural space, the physician slowly injects 10 to 30 ml of a dilute solution (0.25%) of procaine with a soluble steroid into the space. Because this epidural steroid is essentially a sympathetic nerve block, the patient's vital signs need to be monitored for several hours.

In patients who have pain initiated or aggravated by excessive lordosis, especially with lateral flexion and with lateral pain localization, the facet is implicated as a source of pain. Diagnostically, then therapeutically, the facet can be injected with a soluble steroid in an anesthetic agent. This facet injection needs to be performed under fluoroscopic guidance.

When there are sciatic radicular symptoms with a specific nerve root involved, a local nerve block may be valuable (Fig. 3-65). This injection also requires fluoroscopic visualization and careful aspiration to ascertain that the dural sleeve has not been invaded.

SCIATIC NEUROPATHY

Once leg pain, with or without low back pain, has been confirmed as emanating from the nerve root, the pain can be localized (see Figs. 3-43 and 3-44) by eliciting the following:

1. Positive straight leg raising (usually one leg) (see Fig. 3-42) with a positive "dural sign" (aggravation of leg pain by nuchal flexion or ankle dorsiflexion).
2. Subjective and/or objective hypalgesia of a dermatomal area (see Fig. 3-45).

3. Subjective and/or objective evidence of specific muscle weakness or fatigue of a precise myotome (Table 3–2; see Table 3–1).
4. Limited trunk flexion and/or functional scoliosis. This sign occurs often but is not mandatory to establish a root entrapment.

Any or all of these signs may indicate herniation of a specific lumbar disk. The history and a careful orthopedic and neurologic examination confirm the tentative diagnosis. If any doubt exists regarding the etiology, the precise root level, or whether surgical consultation is indicated, diagnostic studies such as computerized tomography (CT), magnetic resonance imaging (MRI), myelography, and EMG studies are indicated.

Treatment for sciatic neuropathy from a mechanical cause is similar to that for acute low back pain without root symptoms. However, bed rest may be of longer duration, traction may be instituted earlier and longer, steroids may be used sooner, and so forth. Even manipulation performed

Table 3–2. RELATIONSHIP OF SPECIFIC ROOTS, MUSCLES, AND PERIPHERAL NERVES

Root	Muscle	Peripheral Nerve
L_2	Sartorius (L_{2-3})	Femoral
	Pectineus (L_{2-3})	Obturator
	Adductor longus (L_{2-3})	Obturator
L_3	Quadriceps femoris (L_{2-3-4})	Femoral
L_4	Quadriceps femoris (L_{2-3-4})	Femoral
	Tensor fascia lata (L_{4-5})	Superior gluteal
	Tibialis anterior (L_{4-5})	Peroneal
L_5	Gluteus medius (L_{4-5} S_1)	Superior gluteal
	Semimembranosus (L_{4-5} S_1)	Sciatic
	Semitendinosus (L_{4-5} S_1)	Sciatic
	Extensor hallucis longis (L_{4-5} S_1)	Deep peroneal
S_1	Gluteus maximus (L_{4-5} S_{1-2})	Inferior gluteal
	Biceps femoris—short head (L_5 S_{1-2})	Sciatic
	Semitendinosus (L_{4-5} S_1)	Sciatic
	Medial gastrocnemius (S_{1-2})	Tibial
	Soleus (S_{1-2})	Tibial
S_2	Biceps femoris—long head (S_{1-2})	Sciatic
	Lateral gastrocnemius (S_{1-2})	Tibial
	Soleus (S_{1-2})	Tibial

carefully may be beneficial if closely monitored by periodic neurologic examinations.

Surgical intervention is indicated if:

1. There is evidence or suspicion of a neurogenic bladder. This is an *emergency* situation, and early intervention is necessary. Significant pressure on the sacral nerves to the bladder cannot be tolerated for any length of time. Hours, rather than days or weeks, are significant in this emergency. Persistence of bladder symptoms after merely 2 to 3 days of acute bladder nerve entrapment have been reported in the literature.
2. Progressive *objective* neurologic findings exist in spite of adequate conservative management.
3. Persistent severe radicular pain remains disabling in spite of conservative treatment *in the presence of definite objective findings: clinical and radiologic.* If the physician has any degree of concern, a psychologic evaluation should precede the surgery to assure that "successful surgery" will remove or decrease the patient's pain.

Surgical *exploration* to locate the cause of pain when *all* the objective findings are negative is to be condemned, as it is the major cause of "failed surgery."

DEGENERATIVE ARTHRITIS (Disk Degeneration)

Inevitably, all humans develop a degree of disk degeneration, predominantly at the L_4-L_5 and L_5-S_1 spaces. Narrowing of the disk space, with or without osteoarthritic spurs, is apparent in x-ray studies of most people after age 50 to 55, yet such degeneration may not be associated with symptoms. Finding these x-ray changes in symptomatic patients does not necessarily alter the therapeutic regimen or the prognosis for low back pain.

SPINAL STENOSIS

Aging of the spinal column in elderly patients often causes signs and symptoms of spinal stenosis.

The spinal canal contains all the nerve roots of the cauda equina with their dural sheaths and their intrinsic vascular supply. If the canal becomes narrowed (the condition termed **stenosis**), these roots are compressed and become relatively ischemic. The cause of this stenosis may be one of the following:

1. Central herniation of the disk into the canal
2. Thickening of the ligamentum flavum (usually associated with disk degeneration)
3. Osteoarthritic zygapophyseal joints with spur formation
4. Congenital stenosis aggravated by any of the previous factors
5. Spondylolisthesis

The symptoms of spinal stenosis are those of pseudoclaudication. These signs and symptoms are pathognomonic and classic:

1. Pain in the lower back and/or of sciatic root distribution brought on by a period of walking or standing. The distance of walking or time of standing before the onset of pain slowly diminishes until significant restriction of the patient's normal activities becomes necessary.
2. Relief of the pain upon flexion of the spine—that is, sitting, bending over, or laying with knees to chest.
3. Initiation of pain, and often of objective signs, by enforced walking.
4. Confirmation of the stenosis by appropriate tests: CT scans, myelography, or MRI.

Conservative treatment for stenosis requires elongation of the spine by decreasing the lordosis and initiating a lumbar kyphosis. This acquired flexed position decreases the stenosis and "unkinks" the enclosed cauda equina.

Exercises to "tilt" the pelvis and "flatten" the lower back, combined with use of a corset and then modification of daily activities, are often valuable. Progression of symptoms in which the impairment markedly decreases the distance of walking and the time of standing without pain justifies surgical decompression.

SPONDYLOLISTHESIS

Spondylolisthesis is a sliding of one vertebra upon the adjacent vertebra. For example, L_4 may slide upon L_5 or L_5 upon S_1. The superior vertebra usually slides anteriorly, but may slide posteriorly (retrospondylolisthesis). This condition is clinically suspected and radiologically confirmed. The history may be exclusively low back pain or may include leg root symptoms. X-ray studies verify the degree of listhesis and the associated mechanism (e.g., lysis or disk degeneration) (Fig. 3–66).

Many patients with spondylolisthesis who are between the ages of 20 and 50—when they remain physically active, are in good condition, and

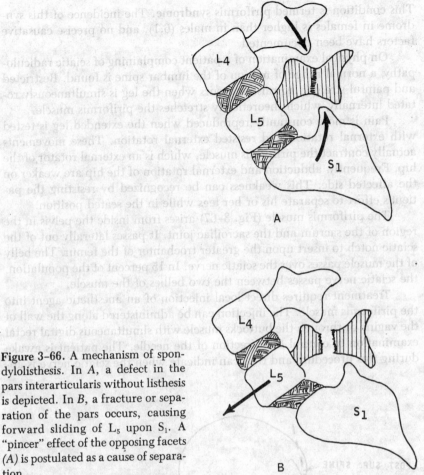

Figure 3–66. A mechanism of spondylolisthesis. In *A*, a defect in the pars interarticularis without listhesis is depicted. In *B*, a fracture or separation of the pars occurs, causing forward sliding of L$_5$ upon S$_1$. A "pincer" effect of the opposing facets (*A*) is postulated as a cause of separation.

use proper body mechanics—remain asymptomatic. Therefore, a conservative approach to treatment is valid. In such an approach, exercises that "flatten" the lower back and a daily activity program are instituted to decrease the lumbosacral angle, strengthen the abdominal muscles, strengthen the extensor muscles of the back, and train the patient to use proper form in bending and lifting activities.

PIRIFORMIS SYNDROME

Low back pain with sciatic radiation may be caused by entrapment of the sciatic nerve as it emerges from under the piriform (piriformis) muscle.

This condition is termed **piriformis syndrome.** The incidence of this syndrome in females is higher than in males (6:1), and no precise causative factors have been documented.

On physical examination of a patient complaining of sciatic radiculopathy, a normal range of motion of the lumbar spine is found. Restricted and painful straight leg raising occurs when the leg is simultaneously rotated internally, which theoretically stretches the piriformis muscle.

Pain is more constantly reproduced when the extended leg is tested with external rotation and resisted external rotation. These movements actually contract the piriformis muscle, which is an external rotator of the hip. Frequently, abduction and external rotation of the hip are weaker on the affected side. This weakness can be recognized by resisting the patient's effort to separate his or her legs while in the seated position.

The piriformis muscle (Fig. 3–67) arises from inside the pelvis in the region of the sacrum and the sacroiliac joint. It passes laterally out of the sciatic notch to insert upon the greater trochanter of the femur. The belly of the muscle passes over the sciatic nerve. In 15 percent of the population, the sciatic nerve passes between the two bellies of the muscle.

Treatment requires direct local injection of an anesthetic agent into the piriformis muscle. The injection can be administered along the wall of the vagina or through the buttocks muscle with simultaneous digital rectal examination to control the direction of the needle. The patient is awake during this procedure and thus can indicate when the sciatic nerve is ap-

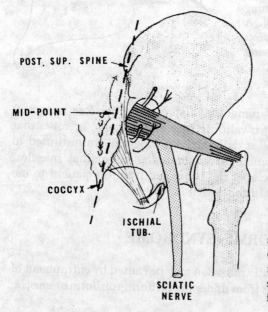

POST. SUP. SPINE

MID-POINT

COCCYX

ISCHIAL TUB.

SCIATIC NERVE

Figure 3–67. Piriform muscle. The sciatic nerve leaves the pelvis along with the piriform muscle through the greater sciatic foramen. The sciatic nerve emerges below the piriform muscle midway between the ischial tuberosity and the greater trochanter. The emergence of the piriform muscle is midway between the posterosuperior iliac spine and the tip of the coccyx. These landmarks aid in determining the site of injection into the piriform muscle.

proached, before the anesthetic agent is injected, to avoid a sciatic chemical neuropathy.

Piriformis syndrome is an infrequent cause of neuropathy, but should be suspected in a female patient with normal lumbar spinal movement who complains of dyspareunia and has sciatic pain reproduced by straight leg raising while the leg is internally rotated. There are no specific tests to confirm this diagnosis.

SUMMARY

Measurement of pain and of human reaction to pain remains the greatest challenge in medical clinical practice. The need for objective documentation is a cause of frustration to the clinician, who must give explanations to third parties, insurance companies, and attorneys regarding a patient's subjective symptom of pain versus the organic or "objective" findings. Injustice may be done to any or all persons involved because of this need to differentiate the subjective pain complaint from the documented objective findings.

Chronic pain localized in the lumbosacral area unquestionably has emotional and psychologic components, and the physician is responsible for evaluating the extent of these emotional factors. Terms that are considered diagnostic enter the report and in turn may influence both the patient's pain and the prescribed treatment. Failure of the patient to respond favorably to "accepted" treatment further clouds the veracity of the patient's report.

The mechanism and anatomic site of the low back pain can usually be discerned from a carefully undertaken history, precisely interpreted physical examination, and properly evaluated laboratory examinations. The patient's pain, however, is an emotional and psychologic response to the sensory input from the injured tissues. To deny that pain is a psychologic response is untenable in light of current medical knowledge.

"Secondary gains" from pain in the lumbosacral area have been clinically documented and voluminously discussed in the literature. Full discussion of these gains is beyond the scope of this chapter, but their existence must be acknowledged. Gains may be monetary, social, or personal, but without question, they are psychologic. This does not mean that pain does not exist: it does. This does not mean that patients are faking or fabricating their disabilities: they are not! The secondary gains must be fully evaluated in the physician's assessment of the disability.

No patient having chronic or recurrent low back pain involving disability—that is, impaired social, vocational, or psychologic function— can be considered to be thoroughly evaluated and competently treated until the psychologic aspect is adequately understood.

The "sick" person—here, the patient with disabling low back pain—enjoys or benefits from a socially accepted impairment and is freed from daily activities. This sick person is granted freedom from total self-care and dependency on others becomes permissible. Socially, a person is considered "sick" when someone in authority labels him and treats him so. The person in authority in today's society is the physician, who thus has an awesome responsibility.

Pain and disability that bring rewards understandably tend to persist, whereas pain and disability that are unrewarded lose their value. This principle is the basis for the current approach to treatment termed **operant conditioning**, and for the financial rewards of legal compensation.

Various treatment modalities are being carefully scrutinized and found to be lacking in objective documentation, yet they persist because of their subjective benefit to patients. The term placebo is constantly used in describing the value of "useless" treatment, whether it is medicine or some other modality. "Placebo" implies that no acceptable objective benefit is derived from that treatment but that the patient accepts a benefit, which is the loss or the diminution of the disabling pain. The fact that the patient benefits from the placebo must be accepted and documented by the physician.

A careful history taken by an attentive physician is the beginning placebo. The "laying on of hands" in the examination is also beneficial as a placebo. Receiving an understandable explanation by the physician of the "results" of a test is a further placebo. Judicious use of medicine with an appropriate explanation of the rationale for, and expected results of, that medicine is a further placebo. Psychologists have recently claimed that the value of many types of surgery is placebo in addition to the eradication of the structural causes of pain. This concept may help explain why so many "proper" types of surgery fail to relieve patient's symptoms.

The existence of and benefit from placebo treatment must be accepted. Such acceptance of a placebo effect does not mean that a person is manufacturing their symptoms or that all symptoms are imaginary. The value of placebo treatment enforces the fact that a careful history, meaningful and appropriate examination, appropriate tests, and proper treatment benefit most patients with low back pain. Each significant aspect of the history and the physical findings must be shared with the patient, and each must be explained.

Eighty to 85 percent of all patients with acute low back pain recover within 3 days to 3 weeks, *regardless of treatment*. The value of treatment, therefore, is to shorten the recovery period and prevent recurrence and progression into chronicity. The purpose of proper evaluation of the patient with low back pain is to uncover "serious organic disease" such as malignancy or metabolic disease. The diagram on the facing page summarizes appropriate patient care as outlined in this chapter.

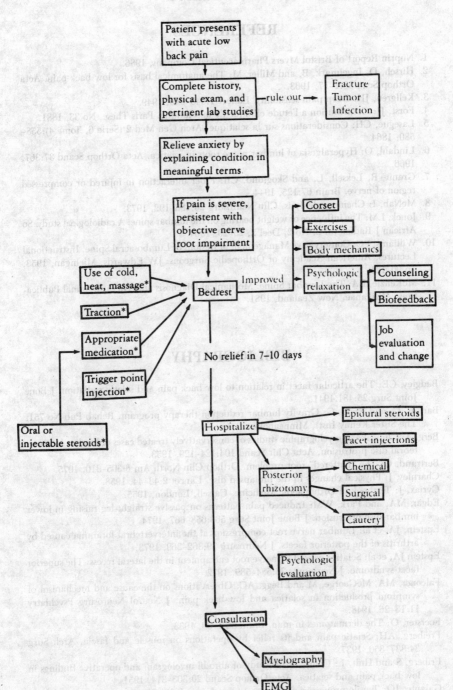

Patient presents with acute low back pain

Complete history, physical exam, and pertinent lab studies — rule out → Fracture / Tumor / Infection

Relieve anxiety by explaining condition in meaningful terms

If pain is severe, persistent with objective nerve root impairment

Corset
Exercises
Body mechanics

Use of cold, heat, massage*
Traction*
Appropriate medication*
Trigger point injection*

Oral or injectable steroids*

Bedrest — Improved → Psychologic relaxation

Counseling
Biofeedback
Job evaluation and change

No relief in 7–10 days

Hospitalize

Epidural steroids
Facet injections

Posterior rhizotomy

Chemical
Surgical
Cautery

Psychologic evaluation

Consultation

Myelography
EMG

*May be used in hospital

REFERENCES

1. Nuprin Report of Bristol Myers Pharmaceutical Company, 1986.
2. Hirsch, D, Ingelmark, B, and Miller, M: The anatomical basis for low back pain. Acta Orthop Scan 33:1–17, 1963.
3. Kellgren, JH: Deep pain sensibility. Lancet 1:943–949, 1949.
4. Forst, JJ: Contribution a l'etude clinique de la sciatique. Paris These, No 33, 1881.
5. Lasegue, CH: Considerations sur la sciatique. Arch Gen Med 2 (Serie 6, Tome 4):558–580, 1864.
6. Lindahl, O: Hyperalgesia of lumbar nerve roots in sciatica. Acta Orthop Scand 37:367, 1966.
7. Granit, R, Leksell, L, and Skoglund, CR: Fiber interaction in injured or compressed region of nerve. Brain 67:125, 1944.
8. McNab, I: Chemonucleolysis. Clin Neurosurg 20:183–192, 1973.
9. Jonck, LM: The influence of weight bearing on the lumbar spine: A radiological study. So African J Radiol Vol 2, No 2, Deel 2, 25–29, 1964.
10. Williams, PC: Conservative Management of Lesions of Lumbosacral Spine. Instructional Lectures, American Academy of Orthopedic Surgeons. JW Edwards, Michigan, 1953, pp 90–121.
11. McKenzie, RA: The Lumbar Spine: Mechanical Diagnosis and Therapy. Spinal Publications, Waikanae, New Zealand, 1981.

BIBLIOGRAPHY

Badgley, CE: The articular facet in relation to low back pain and sciatic radiation. J Bone Joint Surg 25:481,1941.
Barton, C, and Neda, G: Gravity lumbar reduction therapy program. Rehab Pub No 751, The Sister Kenny Inst., Minneapolis, 1976.
Berg, A: Clinical and myelographic studies of conservatively treated cases of lumbar intervertebral disc protrusion. Acta Chir Scand 104:124–129, 1953.
Bertrand, G: The "battered" root problem. Orthop Clin North Am 6:305–310, 1975.
Charnley, J: Physical changes in the prolapsed disc. Lancet 2:43–44, 1958.
Cyriax, J: Textbook of Orthopaedic Medicine. Cassell, London, 1955.
Edgar, MA, and Park, WM: Induced pain patterns on passive straight-leg raising in lower lumbar disc protrusion. J Bone Joint Surg 563:658–667, 1974.
Epstein, JA, et al: Lumbar nerve root compression at the intervertebral foramina caused by arthritis of the posterior facets. J Neurosurg 39:362–369, 1973.
Epstein JA, et al: Sciatica caused by nerve root entrapment in the lateral recess: The superior facet syndrome. J Neurosurg 36:584–589, 1972.
Falconer, MA, McGeorge, M and Begg, AC: Observations on the cause and mechanism of symptom production in sciatica and low-back pain. J Neurol Neurosurg Psychiatry 11:13–26, 1948.
Foerster, O: The dermatomes in man. Brain 56:1–39, 1933.
Freiberg, AH: Sciatic pain and its relief by operations on muscle and fascia. Arch Surg 34:337–350, 1937.
Friberg, S and Hult, L: Comparative study of abrodil myelogram and operative findings in low back pain and sciatica. Acta Orthop Scand 20:303–314, 1951.
Galante, JO: Tensile properties of the human lumbar annulus fibrosus. Acta Orthop Scand Supplement 100, 1967.

Gasser, HS, and Erlanger, J: Role of fiber size in establishment of nerve block by pressure or cocaine. Am J. Physiol 88:581, 1929.

Ghormley, RK: Low back pain with special reference to the articular facets with presentation of an operative procedure. JAMA 101:1773–1776, 1933.

Green, LN: Dexamethasone in the management of symptoms due to herniated lumbar disc. J Neurol Neurosurg Psychiatry 38:1211–1217, 1975.

Hallen, LG: The collagen and ground substance of human intervertebral disc at various ages. Acta Chem Scand 16:705–710, 1962.

Harris, RI, and MacNab, I: Structural changes in the lumbar intervertebral discs. J Bone Joint Surg. 36[BR]:304–322, 1954.

Hartman, JT, et al: Intradural and extradural corticosteroids for sciatic pain. Orthop Review 3:21–24, 1974.

Hendry, NGC: The hydration of the nucleus pulposus and its relation to intervertebral disc derangement. J Bone Joint Surg 40[Br]:132–144, 1958.

Hockaday, JM, and Whitty, CWM: Patterns of referred pain in normal subject. Brain 90:481–496, 1967.

Inman, VT, and Saunders, JB deCM: Referred pain from skeletal structures. J Nerv Ment Dis 99:660–667, 1944.

Inman, VT, and Saunders, JB deCM: The clinico-anatomical aspects of the lumbosacral region. Radiology 38:669–687, 1942.

Inman, VT, et al: Referred pain from experimental irritative lesions. In: Studies Relating to Pain in the Amputee. Series II, Issue 23, pp 49–78, June 1952.

Kelly, M: Is pain due to pressure on nerves? Spinal tissues and the intervertebral disc. Neurology 6:32, 1956.

Krempen, JF, Smith, BS, and DeFreest, LJ: Selective nerve root infiltration for the evaluation of sciatica. Orthop Clin North Am 6:311–315, 1975.

Landahl, C, and Rexed, B: Histological changes in spinal nerve roots of operated cases of sciatica. Acta Orthop Scand 20:215–225, 1951.

Lindblom, K: Technique and results of diagnostic disc puncture and injection (discography) in lumbar region. Acta Orthop Scand 20:315–326, 1951.

Maigne, R: Medical Orthopedics. Charles C Thomas, Springfield, IL, 1975.

Maroudas, A, et al: Factors involved in the nutrition of the human lumbar intervertebral disc: Cellularity and diffusion of glucose in vitro. J Anat 120:113–130, 1975.

Marshall, LL and Trethewie, ER: Chemical irritation of nerve root in disc prolapse. Lancet 2:320, 1973.

McCollum, DE, and Stephen, CR: Use of graduated spinal anesthesia in the differential diagnoses of pain of the back and lower extremities. South Med J 57:410, 1967.

Nachemson, A, et al: In vitro diffusion of dye through the end plates and the annulus fibrosus of human lumbar intervertebral disc. Acta Orthop Scand 41:589–607, 1970.

Pace, JB, and Nagle, D: Piriform syndrome. West J Med 124:435–439, 1976.

Pedersen, HE, Blunch, CEJ, and Gardner, E: The anatomy of lumbosacral posterior rami and meningeal branches of the spinal nerves (sinu-vertebral nerves). J Bone Joint Surg 38[Am]:377–390, 1956.

Perey, O: Contrast medium examination of intervertebral discs of lower lumbar spine. Acta Orthop Scand 20:327–334, 1951.

Raaf, J: Some observations regarding 905 patients operated upon for protruded lumbar intervertebral disc. Am J Surg 97:388–397, 1980.

Rees, WES: Multiple bilateral subcutaneous rhizolysis of segmental nerves on the treatment of the intervertebral disc syndrome. Ann Genet 26:126–127, 1971.

Sarnoff, SJ, and Arrowwood, JG: Differential spinal block. Surgery 20:150, 1946.

Sehgal, AD, et al: Laboratory studies after intrathecal corticosteroids. Arch Neurol 9:64–68, 1963.

Shealy, CN: Facets in back and sciatic pain: A new approach to a major pain syndrome. Minn Med 57:199–203, 1974.

Sinclair, DC, et al: Intervertebral ligaments as source of segmental pain. J Bone Joint Surg 30[Br]:515–521, 1948.

Smyth, MJ, and Wright, V: Sciatica and the intervertebral disc. An experimental study. J Bone Joint Surg 40[Am]:1401, 1958.

Sylven, B: On biology of nucleus polposus. Acta Orthop Scand 20:275–279, 1951.

Troop, JDG: Ph.D. Thesis, London University, 1968.

Winnie, AP, and Collins, VJ: Pain Clinic, I. Differential neural blockade in pain syndromes of questionable etiology. Med Clin North Am 52:123–129, 1968.

Winnie, AP, et al: Pain Clinic, II. Intradural and extradural corticosteroids for sciatica. Anesth Analg 51:990–999, 1973.

CHAPTER 4

Neck and Upper Arm Pain

As a medical complaint, pain in the neck, the head, or the interscapular area (referred from the neck) is second in frequency only to low back pain.

Simplistically stated, there are three major musculoskeletal causes of neck pain or of upper extremity pain referred from the neck: "trauma," posture, and tension. Obviously, conditions more serious than musculoskeletal injury, such as inflammatory arthritis, fracture, dislocation, or metastatic lesions, must always be considered and ruled out by appropriate tests.

As the vast majority of complaints of neck pain have the three mentioned causes, the physician must establish what happened to the neck, what alteration of normal neuromusculoskeletal anatomy has occurred, and what tissues have been irritated and inflamed.

FUNCTIONAL ANATOMY

A complete knowledge of the functional anatomy of the cervical spine is needed to interpret the history correctly and to perform a meaningful examination. Appropriate tests then follow to confirm the suspected lesion. Appropriate treatment necessarily implies restoring normal function as completely as possible.

Functional anatomy has been fully discussed in my book *Neck and Arm Pain*,[1] but a summary of pertinent functional anatomy and appropriate neurophysiologic aspects merits repetition here.

For clinical purposes, the cervical spine can be seen as having an up-

123

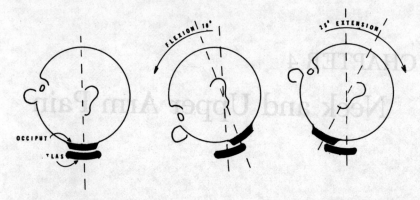

Figure 4–1. Motion at the occipito-atlas joint, permitting 10 to 20 degrees flexion and 25 degrees extension, totaling 35 to 40 degrees of flexion-extension. No rotation is possible.

per and a lower segment. These segments are anatomically and functionally different and respond differently to injury.

The upper cervical segment contains the occipitocervical (cervicoatlas) articulation, the atlanto-axis (C_1–C_2) articulation, and the axocervical (C_2–C_3) functional units. The lower cervical spinal segment includes the remaining cervical spine (C_3–C_8) functional units. Each of these lower functional units are essentially similar in structure and in function.

At the occipito-atlas articulation, only flexion and extension occur. These motions are 10 degrees of flexion and 25 degrees of extension from a neutral position (Fig. 4–1).

Between the atlas and the axis (C_1–C_2), rotation occurs about the odontoid process of the axis (C_2); approximately 45 degrees of rotation to each side, a total of 90 degrees rotation, is permitted (Fig. 4–2). A slight degree of flexion and extension is permitted at this C_1–C_2 articulation. Without motion occurring elsewhere in the cervical spine, the upper cervi-

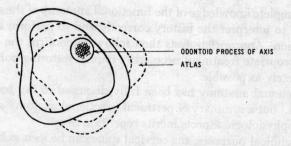

ODONTOID PROCESS OF AXIS
ATLAS

Figure 4–2. Motion of the atlas about the odontoid process of the axis, permitting rotation of 75 to 90 degrees.

Figure 4-3. Locking mechanism of C₂ upon C₃. Rotation of C₂ upon C₃ is limited by the mechanical locking of the articular structures. The anterior tip of the upper articular process of C₃ (F) impinges upon the lateral margin of the foramen of the vertebral artery (V). G is the gutter through which emerges the nerve root C₃.

cal segment, occiput to C_2, can move with 45 degrees of flexion-extension and 90 degrees of rotation.

Rotation of the second cervical vertebra upon the third cervical vertebra is limited by a bony locking mechanism, in which the anterior tip of the upper articular process of the third cervical vertebra impinges upon the lateral process of the second vertebra (Fig. 4-3). This locking mechanism prevents excessive rotation and thus protects the vertebral artery and the nerve root, which descend the spinal nerve groove.

No intervertebral disks exist at the occipito-atlas articulations or the atlas-axis articulations, and the posterior intervertebral foramina are *not* delineated by structures found in the functional units of the lower segment. These structures are the pedicles, disks, and the posterior zygapophyseal joints (Fig. 4-4).

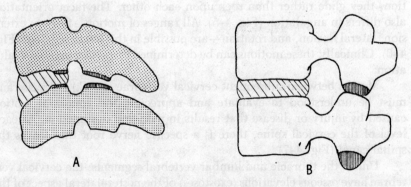

Figure 4-4. Comparative lateral views of cervical and lumbar functional units. *A*, Curved vertebral bodies of the cervical spine, the joints of Luschka *(shown in the stippling)*, and the intervertebral foramina. *B*, Lumbar vertebrae with different vertebral body contours and no joints of Luschka.

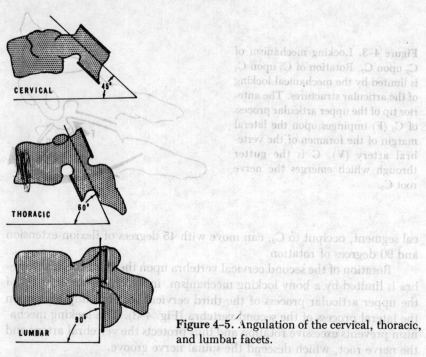

Figure 4-5. Angulation of the cervical, thoracic, and lumbar facets.

The stability of the atlanto-axial articulation is maintained by ligamentous support. This ligamentous support becomes weakened in rheumatoid disease.

The cervical vertebrae of the lower segment, unlike their thoracic and lumbar counterparts, have more concavity and convexity, and in their motion, they glide rather than rock upon each other. The facet orientation also differs in angulation (Fig. 4-5). All ranges of motion—flexion, extension, lateral flexion, and rotation—are possible in this lower segment (Fig. 4-6). Clinically, these motions can be determined by precise manual evaluation.

Motion between subsequent cervical vertebrae (C_3–C_8) is precise and must be understood to evaluate and appreciate the abnormal motion caused by injury or disease that results in pain and dysfunction. At each level of the cervical spine, there is a specific nerve root that leaves the spinal canal (Fig. 4-7).

Unlike the thoracic and lumbar vertebral segments, the cervical vertebrae have osseous elevations (exostoses) of the posterolateral aspect of the vertebral bodies that form pseudojoints, which are termed uncovertebral joints (Fig. 4-8). The joints are positioned so that they protect the contents of the spinal canal from intervertebral disk protrusion. They pose a pathologic problem in that they undergo hypertrophy and further calcification

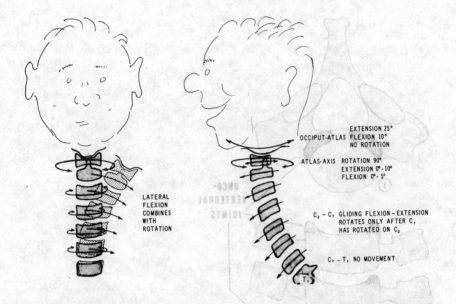

Figure 4–6. Composite movements of the cervical spine with occipito-atlas flexion-extension but no rotation. There is 90 percent of rotation at the axoatlantoid joint and further movement of C_2 through C_7.

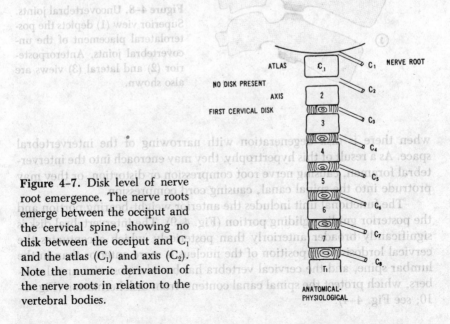

Figure 4–7. Disk level of nerve root emergence. The nerve roots emerge between the occiput and the cervical spine, showing no disk between the occiput and C_1 and the atlas (C_1) and axis (C_2). Note the numeric derivation of the nerve roots in relation to the vertebral bodies.

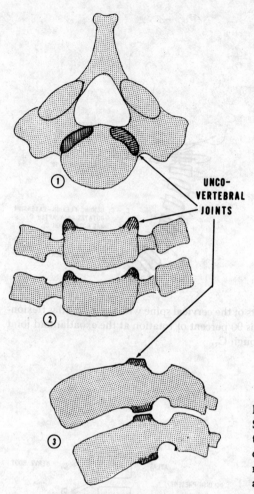

UNCO-
VERTEBRAL
JOINTS

Figure 4-8. Uncovertebral joints. Superior view (1) depicts the posterolateral placement of the uncovertebral joints. Anterorposterior (2) and lateral (3) views are also shown.

when there is disk degeneration with narrowing of the intervertebral space. As a result of this hypertrophy, they may encroach into the intervertebral foramen, causing nerve root compression or distortion, or they may protrude into the spinal canal, causing cord compression.

The functional unit includes the anterior weight-bearing portion and the posterior guiding-gliding portion (Fig. 4-9). The intervertebral disk is significantly broader anteriorly than posteriorly, which accentuates the cervical lordosis. The position of the nucleus is more anterior than in the lumbar spine, and the cervical vertebra has broader posterior annular fibers, which protect the spinal canal contents from disk herniation (Fig. 4-10; see Fig. 4-4).

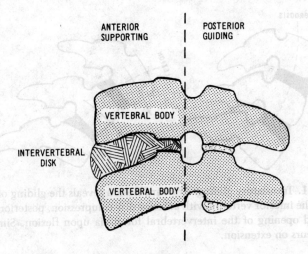

Figure 4-9. The functional unit in cross section.

The musculature of the cervical spine is adequately clarified in most textbooks of anatomy and kinesiology. I want to note here, however, that the extensor muscles are the first to develop and remain the predominant muscles throughout life. In neonatal life, the cervical extensor muscles develop early and are instrumental in forming the cervical lordosis. Until the individual assumes the erect posture, these extensor muscles support the head against the forces of gravity. When the erect posture is assumed, the

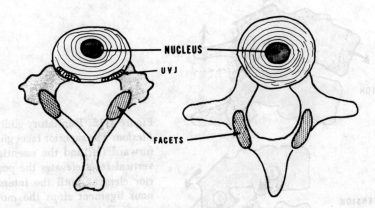

Figure 4-10. Comparison of cervical and lumbar vertebrae. *Left,* Cervical vertebra with uncovertebral joints (UVJ), anteriorly placed nucleus, and broader posterior annular fibers. *Right,* Lumbar vertebra with a centrally placed nucleus and no uncovertebral joints. The angulation of the facets also differs as indicated.

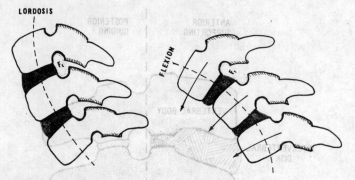

Figure 4–11. Translatory gliding. Forward flexion reveals the gliding of the superior upon the inferior vertebra with anterior disk compression, posterior disk separation, and opening of the intervertebral foramina upon flexion. Simultaneous closure occurs on extension.

erector spinae muscles maintain the head's erect position and permit the neck to flex forward by elongation and eccentric contraction. The flexors of the neck function primarily when force is applied to the neck to push the head backward, or when gravity must be overcome, such as in lifting the head from the supine position.

Motion is that of translatory gliding (Fig. 4–11). The disk deforms proportionately to allow this motion. Since the cervical facets are at a 45-degree angle with the foraminal plane, any forward motion (flexion)

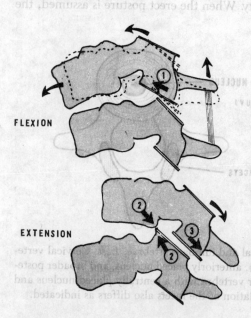

FLEXION

EXTENSION

Figure 4–12. Translatory gliding. *Flexion:* The superior facet glides forward (1), and the essentially vertical facet elevates the posterior element until the interspinous ligament stops the movement. *Extension:* The superior facet glides posteriorly (3) until the inferior facet impinges upon the vertebra (2) and locks further extension movement.

causes the facets to glide upon each other. Thus, the posterior elements are elevated until, ultimately, the interspinous ligaments are stretched to their maximum. At this point, further flexion is stopped (Fig. 4–12; see Fig. 4–10).

The nuchal ligament is an elastic ligament consisting largely of parallel collagen fibers with a variable amount of elastin fibers. This tissue is essentially avascular, but is abundantly supplied by nerves, which arise from the posterior primary division of the second, third, and fourth cervical nerve roots.

Recent experiments have demonstrated that these nerves have proprioceptive receptors and act in attitudinal and postural tonic neck reflexes. This finding adds credence to the many bizarre neurologic symptoms of cervical spine injuries.

Excessive flexion force imposed upon the functional units can cause disruption of the interspinous ligaments and actually result in a subluxation of the facet joints.

Extension of the cervical functional unit causes a posterior gliding of the facets with approximation of the posterior elements. Ultimately, the facets contact each other and prevent further extension. Extension narrows the intervertebral foramina, but the "locking" mechanism, which prevents further extension, prevents compression of the foraminal contents: the nerve roots, dura, and dural contents.

Excessive extension can result in (1) overstretching or even tearing of the restraining anterior longitudinal ligament, (2) fracture of the opposing

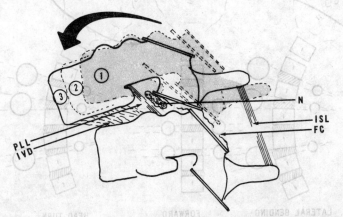

Figure 4–13. Hyperextension-hyperflexion injury. Normal physiologic flexion (1 to 2) is possible with no soft tissue damage. When motion is exceeded (3), the intervertebral disk (IVD) is pathologically deformed, and the posterior longitudinal ligament (PLL) is strained or torn, the nerve (N) is acutely entrapped, the facet capsule (FC) is torn or stretched, and the interspinous ligament (ISL) is damaged.

facets, (3) subluxation of the facets, or (4) acute compression of the contents of the intervertebral foramina (Fig. 4–13).

Lateral motion and rotatory motion occur simultaneously. Neither motion can occur independently, owing to the axis of rotation of the functional unit and the planes of the facets. In this simultaneous motion, the intervertebral foramina open and close: they close on the side toward which lateral flexion and rotation occur, and they open on the side away from lateral flexion and rotation (Fig. 4–14). Lateral flexion to the right closes the foramina on the right and opens those on the left. The reverse

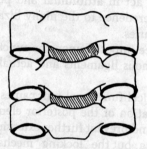

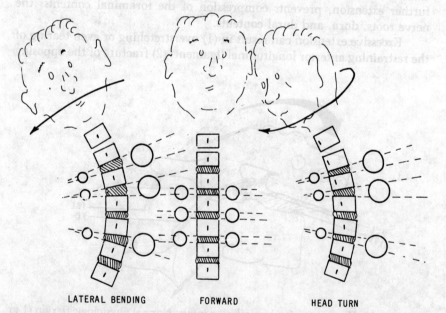

LATERAL BENDING   FORWARD   HEAD TURN

Figure 4–14. Intervertebral foraminal opening and closing on spine motion. Lateral bending closes the intervertebral foramina on the concave side and opens it on the convex side. Rotation, head turning, causes closure on the side toward which rotation occurs, and opening on the side away from which rotation occurs.

occurs in that lateral flexion and rotation to the left closes the foramina on the left and opens those on the right.

A greater degree of facet approximation and foraminal closure occurs when lateral flexion and rotation are combined with simultaneous cervical hyperextension. This fact is clinically significant both in the patient's history and in the examination.

The nerve roots that emerge via the foramina are physiologically prevented from injury during normal movement. Such prevention is assured by the fact that the foramina open and close as described, but also by the simultaneous adjustment of the spinal canal and its contents (the cord and nerve roots).

In flexion, the spinal canal elongates (Fig. 4–15). The contained spinal cord can elongate because it is plastic; the dura can elongate because it is plicated. The nerve roots change their angles of emergence from the cord during these cord and dural changes. The pedicles separate during flexion

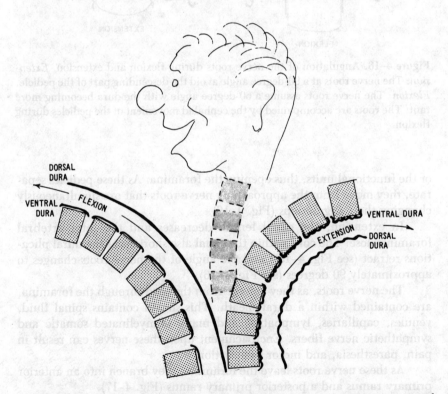

Figure 4–15. Length of spinal canal and dural sheath. Flexion elongates the spinal canal, causing full elongation of the dural sheath. Extension shortens the length of the spinal canal, causing plication of the dural sheath.

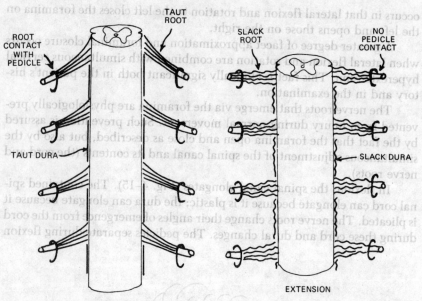

Figure 4–16. Angulation of the nerve roots during flexion and extension. *Extension:* The nerve roots at a 90-degree angle avoid the descending part of the pedicle. *Flexion:* The nerve roots assume a 60-degree angle with the dura becoming more taut. The roots are accompanied by the cephalad movement of the pedicles during flexion.

of the functional units, thus opening the foramina. As these pedicles separate, they move from the approaching nerve roots that are simultaneously changing their angulations (Fig. 4–16).

In extension, the canal length decreases and the intervertebral foramina close. The cord within the canal also shortens as the dural plications retract (see Fig. 4–15), and the angle of the nerve roots changes to approximately 90 degrees (see Fig. 4–16).

The nerve roots, as they emerge from the cord through the foramina, are contained within a dural sheath. This sheath contains spinal fluid, venules, capillaries, lymphatics, and many unmyelinated somatic and sympathetic nerve fibers. Encroachment upon these nerves can result in pain, paresthesia, and motor dysfunction.

As these nerve roots leave the foramina, they branch into an anterior primary ramus and a posterior primary ramus (Fig. 4–17).

Since pain may occur from injury, misuse, or abuse of the cervical spine, the nociceptive tissue sites from which pain may be initiated must be clarified.

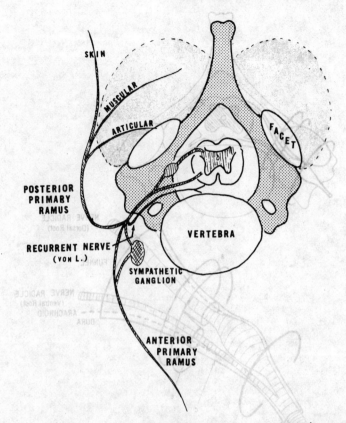

Figure 4–17. Innervation of the cervical roots. The posterior primary ramus divides into skin, muscular, and articular branches; the anterior primary ramus, with a sympathetic innervation, proceeds to the dermatomal areas. (von L.—von Luschka)

The anterior longitudinal ligament is a sensitive tissue. Pain in the vertebral bodies can be elicited from conditions such as metastatic disease, fracture, osteomyelitis, or multiple myeloma, but in such situations, the complaint is of deep nonspecific continuous pain that is, as a rule, unrelated to motion.

The intervertebral disk in the adult is avascular. No nerve capable of transmitting sensation has been found to penetrate the outer annular fibers or the nucleus, so the intervertebral disk has no nerve supply.

The disk's insensitivity to pain remains controversial. According to a study by Cloward,[2] no pain resulted within the intact disk during diskography. Only in the case of degeneration was pain elicited. Holt[3] disagreed, claiming that pain resulted from *any* disk injection.

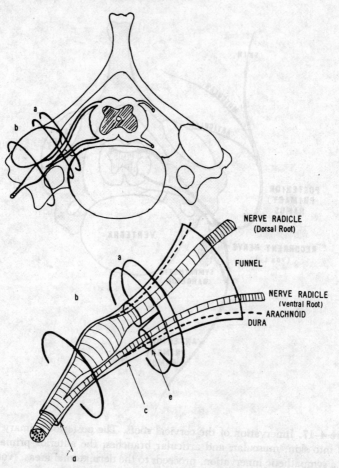

Figure 4-18. Schematic diagram of the sheaths of the dura accompanying the nerve roots in the intervertebral foramina.
a—intervertebral foramen
b—gutter of the transverse process
c—attachment of the dura
d—nerve with only dural sheath
e—apex of funnel

The sensitivity of the anterior longitudinal ligament has been proved. Its nerve supply has been identified as the recurrent nerve of Luschka. Pressure or irritation, such as that produced by a needle, was shown by Cloward[2] to cause referred pain to the interscapular area. Each disk level (C$_4$–C$_5$, C$_5$–C$_6$, and so forth) caused referral of pain to a precise location in the interscapular area. Pressure from lateral bulging of the disk was shown to cause radiation of pain from precise levels of the nerve roots.

The posterior longitudinal ligament is sensitive; it is supplied by the recurrent nerve of Luschka. The nerve root dura within the foramina (Fig. 4–18) is also supplied by the nerve of Luschka and is also a site of nociceptive impulses. The posterior facet joints, supplied by the somatic and sympathetic nerves within the posterior primary division, can become a source of pain when irritated.

The neck muscles, like all muscles in the body, and the ligaments are sites of painful stimuli. Many tissues in the cervical spine are capable of transmitting pain (see Fig. 4–17).

Pain in the neck or referred from the neck must be related to irritation of any of these sensitive tissues. The manner in which this tissue is irritated and causes pain, and the resultant disability, can be elicited from a careful history and functional examination. As the site of tissue injury and the mechanism of the irritation become apparent, a diagnosis evolves, and meaningful treatment is indicated.

Injury to the neck may be classified as acute, recurrent, or chronic. There are three causes of mechanical injury to the cervical spine: (1) external "trauma," (2) posture, and (3) tension. Details of the injury are related by the history. The examination confirms the type and extent of injury— that is, whether it is musculoskeletal and whether it involves neurologic sequelae.

ACUTE TRAUMA

External traumatic forces imposed on the neck may result from an automobile accident, an athletic injury, a fall, or a direct blow. The extent of injury cannot always be immediately determined; thus, the primary care of the injured patient must be carefully monitored and repeatedly evaluated. The severity of the "impact of the injury" is not always equal to the severity of the resultant injury.

At the scene of an accident, one must assume that the patient's neck has been injured, and even that the spinal cord may have been traumatized. These assumptions are especially applicable when the patient has undergone a period of unconsciousness or even amnesia or shock, with or without head trauma or laceration. In a severe or potentially severe injury, the patient must be moved with utmost care and with the neck splinted or immobilized. The head must be kept from any flexion or extension. A slight degree of traction is usually permissible *to hold but not to mobilize* the head and neck. The purpose of any gentle traction is to assure straight alignment and immobilization, not to correct an apparent defect.

X-ray examination, important in an acute injury, should be performed early but carefully, with the patient's head held immobile by the physician or a skilled x-ray technician. The decision regarding which specialist (orthopedist or neurosurgeon) is best for early consultation should be based on the available specialists' experience in spinal trauma.

An adequate airway must be maintained while the patient is still at the scene of the accident. The oral airway requires some hyperextension of the patient's neck for insertion, which is not permitted in the case of fracture or dislocation of the neck with possible cord embarrassment. A nasal latex airway can be simply inserted to assure air passage. No analgesics should be given to the injured patient, because most depress alertness and interfere with the interpretation of pain and vital signs.

Further care of the patient with cervical fracture or dislocation, with or without neurologic deficit, is beyond the scope of this discussion and requires the intervention of a specialist. The precautions listed here are to ensure that the patient reaches this specialist without additional damage or injury.

Hyperflexion-Hyperextension Injury

An injury to the cervical spine that is essentially a hyperflexion-hyperextension injury without major nerve or cord damage may occur. The symptoms are attributed to "soft tissue" injury. The vast majority of these injuries involve external trauma from an automobile accident. Other forms of injury can result from sports activities, stepping into an unexpected hole, or encountering a missing step on a stairway. The mechanisms for these injuries are identical, but as injury from an accident involving two automobiles is so common, it will be discussed as the model for this type of injury.

The person seated in the car may be struck from behind by a moving vehicle. The impact first occurs to the trunk and shoulder, moving them forward. The impacted car decelerates while the immediate cervical movement is flexion of the head upon the neck, the occipitocervical upper segment, then almost instantaneous flexion of the lower cervical segment (C_3–C_8), primarily at C_5–C_6 (Fig. 4–19).

When the speed of the rear car is approximately 15 miles per hour, the impact can often be accepted by the cervical spine. Above 20 miles per hour, injury usually occurs. Certain variables need emphasis. If a person is totally oblivious to the impending impact, the injury is more severe. If aware of the impending impact, the person can "brace" for the impact and the muscles are not "caught off guard." If a person is looking straight ahead, the flexion-extension injury is less than if the person is turned to one side. In the latter rotational position, the impact force is a shear in a lateral rotatory torque motion (Fig. 4–20).

The forward flexion occurs within the first 250 milliseconds (msec): one quarter of a second from the moment of the impact. From inetria, the

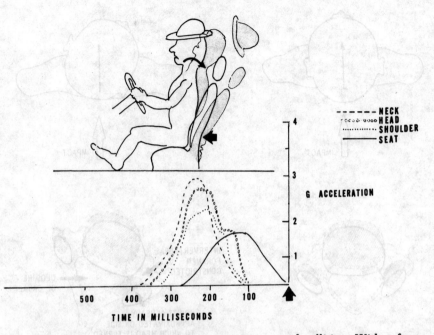

Figure 4-19. Upper body-neck deformation in rear-end collision. With a force from behind, the head, neck, and shoulders deform at different points in time—all happening in less than 0.5 sec.

shoulder and neck, moving forward from the moment of impact, now extend. This occurs within 200 msec with the same sequence of cervical spine movement: occipitocervical then lower cervical movement, with focus on the C_5–C_6 level.

From elastic recoil of the anterior tissues of the cervical spine, the ligaments, the muscles, and the fascia, the head acutely flexes forward. This complicated mechanism of a neck injury sustained from a rear-end collision explains the multiple symptoms that arise from this type of injury.

In injuries with no appreciably severe impact, and no significant subjective neurologic symptoms, an acute injury involving strain or sprain of soft tissue must be assumed to have occurred. The accident need not have been severe in terms of damage to the vehicles to have caused a cervical injury. Incidents such as stepping off a curb of unexpected height, slipping on an icy surface, receiving a blow from behind in an athletic activity, and so forth may also be initial causes of injury.

Each segment of the cervical spine has a physiologic degree and direction of movement. These physiologic ranges may be exceeded by an external force. The movement of the occiput upon the atlas permits little if any rotation or lateral flexion. Movement of C_1–C_2 permits very limited flex-

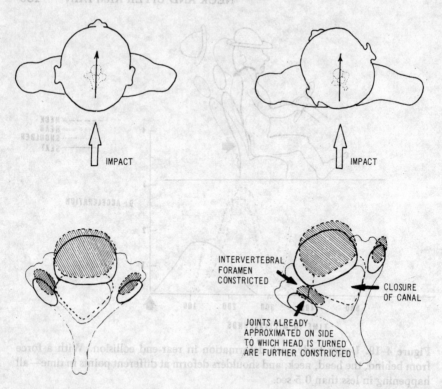

Figure 4-20. Hyperflexion injury with head facing forward versus effect of head rotation. An impact from the rear when head is turned causes further intervertebral foramen closure, excessive spinal canal closure, and excessive joint subluxation. This explains severity of injuries when a car occupant is struck when the head is turned to left or right.

ion, extension, and so forth. External force may cause the body to exceed its normal range of motion. Excessive movement may injure ligaments (the anterior or posterior longitudinal ligaments or those between the lamina or posterior superior spinous processes), nerves and their dura, articular capsules, and muscles. Even the spinal cord and the vertebral arteries may be exposed to trauma.

In considering injuries resulting from hyperextension and hyperflexion, emphasis should be placed on the prefix "hyper" (Fig. 4-21). Normally, the neck can extend until the head, with a line drawn through the occiput to the chin, reaches a 45-degree angle to the vertical plane. If this angle is exceeded, that is, if the angle of the head goes further, the restraining tissues must yield. Subluxation of the involved joints has occurred, and all the soft tissues have, to a degree, been injured.

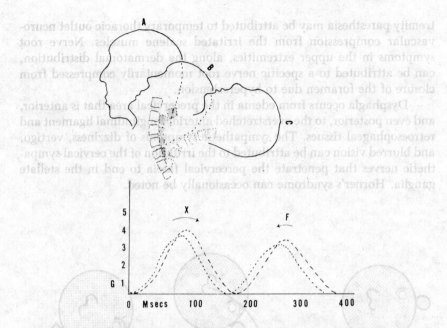

Figure 4-21. Hyperextension injury from rear-end automobile collision. The effects of the accident occur within 300 msec, with the head upon the cervical spine (occipitocervical articulations, A) reacting first (B), followed by extension of the cervical (lower segment) spine (C). X indicates initial flexion, and F indicates reactive flexion.

These injured tissues may be the anterior longitudinal ligaments, the annular fibers of the disk, the facet capsules, the interspinous ligaments, and the cervical muscles. All of these tissues, with the possible exception of the annular disk fibers, are capable of producing nociceptive stimuli that become painful.

Clinical studies and experiments have proved that the brain, contained and floating within the skull, can undergo trauma with resultant edema and even microscopic hemorrhage. The concepts of "concussion" and contrecoup injury are based on this finding, which may explain the temporary loss of consciousness occurring immediately after the accident.

The symptoms of headache, pain between the scapulae, or paresthesia down the arms can be explained by injury to these soft tissues. The headache may be explained by injury to the articular branch of the greater superior occipital nerve, resulting from trauma to the capsule of the C_1-C_2 joint. This topic is discussed more fully in a later section of this chapter (see "Headache").

Cloward[2] has explained the interscapular pain as resulting from injury to the overstretched anterior longitudinal ligament. The upper extremity paresthesia may be attributed to temporary thoracic outlet neuro-vascular compression from the irritated scalene muscles. Nerve root symptoms in the upper extremities, along the dermatomal distribution, can be attributed to a specific nerve root momentarily compressed from closure of the foramen due to hyperextension.

Dysphagia occurs from edema in the precervical area that is anterior, and even posterior, to the overstretched anterior longitudinal ligament and retroesophageal tissues. The sympathetic symptoms of dizziness, vertigo, and blurred vision can be attributed to the irritation of the cervical sympathetic nerves that penetrate the percervical fascia to end in the stellate ganglia. Horner's syndrome can occasionally be noted.

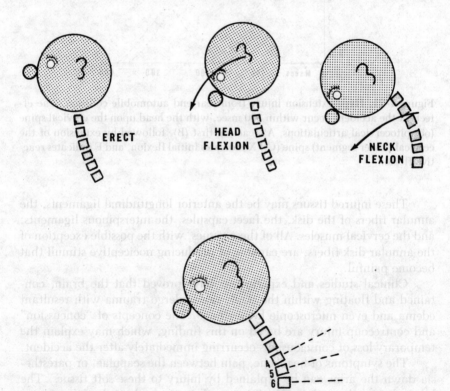

Figure 4–22. Head and cervical spine flexion. In head flexion, the head is flexed upon the cervical spine with movement only at the occipito-atlas articulation; in neck flexion there is reversal of the cervical lordosis. Most flexion occurs between C_4 and C_5 or C_5 and C_6.

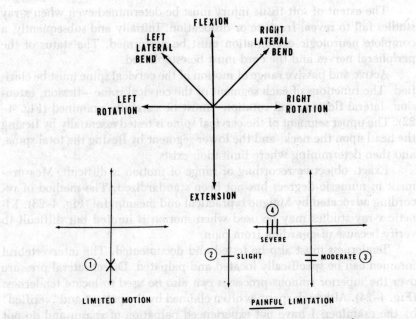

Figure 4-23. Recording of range of motion of cervical spine. X indicates mere limitation and no pain (1), and lines are placed for restriction and simultaneous pain (2 to 4).

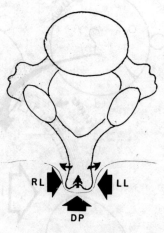

Figure 4-24. Methods of movement of the vertebral segment with direct pressure (DP) or right lateral (RL) or left lateral (LL) pressure upon the posterior superior spine.

Diagnosis and Evaluation

The extent of soft tissue injury must be determined even when x-ray studies fail to reveal fracture or dislocation. Initially and subsequently, a complete neurologic examination must be performed. The status of the peripheral nerves and the cord must be established.

Active and passive range of motion of the cervical spine must be clarified. The functions of each segment of the cervical spine—flexion, extension, lateral flexion, and rotation—must be carefully examined (Fig. 4–22). The upper segment of the cervical spine is tested essentially by flexing the head upon the neck, and the lower segment by flexing the total spine, and then determining where limitation exists.

Exact, objective recording of range of motion is difficult. Measurement in numeric degrees has not been standardized. The method of recording advocated by Maigne is practical and meaningful (Fig. 4–23). Kinetic x-ray studies may be used when motion is limited but difficult to verify because of guarding from pain.

Tenderness must also be tested and documented. The intervertebral foramen can be specifically located and palpated. Direct lateral pressure over the superior spinous processes can also be used to locate tenderness (Fig. 4–24). Although spasm is often claimed by the patient and "verified" by the examiner, I have not experienced palpation of spasm and do not believe that spasm can be palpated. Nodules within a muscle can be pal-

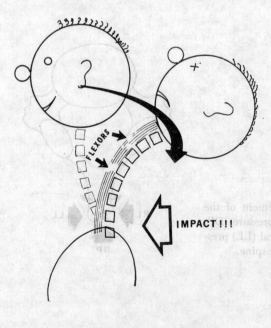

IMPACT!!!

Figure 4–25. Neck flexor trauma from rear-end injury. Hyperextension causes overstretching and inappropriate contraction of the neck flexors with residual flexor disability.

pated, and tenderness elicited, but the presence of spasm is difficult to substantiate.

Owing to overextension (excessive elongation) of the flexor muscles from a hyperextension injury (Fig. 4–25), the patient may complain of difficulty in swallowing, discomfort in the front of the neck, and often, weakness and even total inability to elevate the head from the pillow or the examining table when in the supine position.

Treatment

Pain is the immediate complaint following injury, but anxiety and concern also occur immediately and often predominately. A meaningful examination followed by an explanation of "what happened to the neck," given in understandable terminology, is the first priority of treatment. The patient may regard many of the symptoms as bizarre, and may imagine the injury similar to a chronic disabling condition in a friend or relative who remains seriously impaired. Anger at what happened to the car, frustration caused by the injury's interference with expected activities, the question "why me?" and so forth all compound the patient's reaction to the injury.

As the soft tissue trauma has initiated nociceptive stimuli at the tissue site of injury (see Chapter 2), these stimuli set up the first mechanism of pain and a reflex of muscular contraction termed "spasm."

Splinting and supporting the injured part are valuable. Bed rest with proper pillow position relieves the need for the neck muscles to support the head. This neutral position is best tolerated and avoids placing significant flexion and extension forces upon the injured tissues.

Collar and Brace. A cervical collar or splint has traditionally been considered beneficial in the treatment of neck injury. The purpose of the collar is to hold the head in a comfortable, neutral position and the neck in a neutral, central, mildly lordotic (physiologic) position. The collar is intended to prevent or minimize any movement of the inflamed tissues.

Many types of collars are advocated, but the principles of bracing must be considered in choosing a collar. It must hold the head and neck in the physiologic position.

The collar depicted in Figure 4–26 is made of felt that is $1/4$ to $1/2$ inch thick and contoured to the patient's neck and chin. The felt is enclosed in stockinet, which permits attachment, cleanliness, and comfort. One must understand that at best, this collar restricts merely 5 to 10 percent of flexion, extension, and lateral flexion, as well as some rotation. It functions by allowing the chin to rest upon the anterior portion of the felt, restricting motion by its contact, and furnishing warmth to the neck tissues. It has the

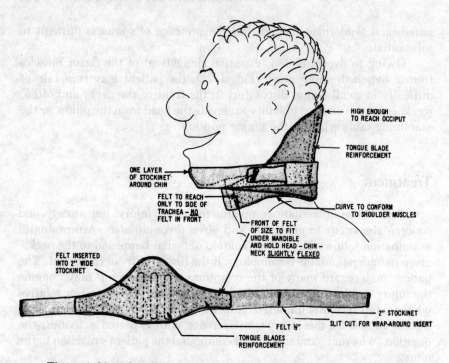

Figure 4–26. Felt collar recommended for occipitocervical restriction.

advantage of being inexpensive and custom made by the patient or therapist.

Plastic collars are firmer than felt collars and have been estimated to restrict movement as much as 50 to 75 percent, but custom molding of these collars is more difficult.

The four-poster cervical brace is cumbersome, and attaining and maintaining the proper adjustment are difficult. This brace restricts as much as a plastic collar does, but it permits atlanto-axial rotation.

More recently, a brace termed the Guildford cervical brace (Fig. 4–27), also termed the suboccipital mandibular immobilizer (SOMI), has been shown to restrict flexion, extension, and rotation by 90 percent. This brace is comfortable, and once properly fitted, can be easily used by the patient.

The only collar or brace proven to immobilize the cervical spine fully is the Halo body jacket brace. Obviously, the choice of the collar or brace is determined by the clinical needs for immobilization.

As a rule, a collar should be worn constantly for the first week, and then half that time for the second week, more often during periods of fatigue or extra activities. Most tissue repair and recovery occur during the

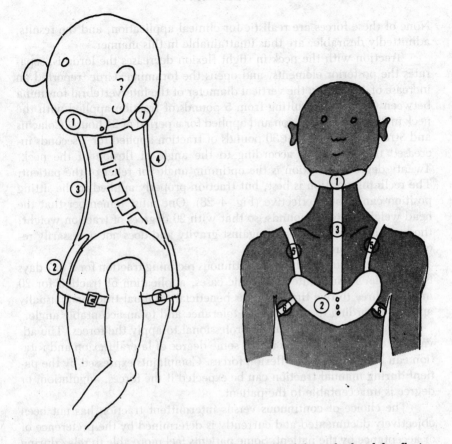

Figure 4–27. Guildford cervical brace. Brace immobilizes the head for flexion-extension and rotation by chin rest (1) and occipital pad (7), which can be narrowed or lengthened to fit. Chest pad (2) is secured to rib cage by shoulder straps (5) and chest straps (6). Upright bars (3) and (4) adjust the flexion and extension of the spine to determine the degree of flexion.

first two weeks. Prolonged immobilization encourages soft tissue contracture, muscle atrophy, and psychologic dependence.

Traction. Like splinting, traction has been well regarded in the treatment of neck injury and possibly has clinical value if applied properly, with its effects objectively evaluated. The best force, direction, method, and duration of traction remain unclarified. De Seze and Levernieux[4] determined that 260 pounds of traction produced approximately 2 mm of separation between the fifth and sixth cervical vertebrae and the sixth and seventh cervical vertebrae. Four hundred pounds of traction produced 10 mm separation between the fourth cervical and first thoracic vertebrae.

None of these forces are realistic for clinical application, and the results, admittedly desirable, are thus unattainable in this manner.

Traction with the neck in slight flexion decreases the lordosis, separates the posterior elements, and opens the foramina. Crue[5] reported an increase of 1.5 mm in the vertical diameter of the intervertebral foramina between C_5 and C_6 resulting from 5 pounds of traction applied with the neck in 20 degrees of flexion and applied for a period of 24 hours. Colachis and Strohm[6] deduced that 30 pounds of traction applied for 7 seconds increased the separation according to the angle of flexion of the neck. Twenty degrees of flexion is the optimum angle for relief of the patient. The reclining position is best, but traction properly applied in the sitting position can also be effective (Fig. 4–28). One must remember that the head weighs 10 to 12 pounds, so that with 20 degrees of traction weight, the head is merely supported against gravity and does not necessarily receive traction forces.

In severe acute problems, continuous reclining traction for 2 to 7 days is desirable. In subacute or chronic cases, application of traction for 20 minutes three to five times daily is beneficial. Manual traction is usually applied according to the patient's tolerance and to an acceptable angle.

Manual traction requires a professional to apply the forces. The advantage of manual traction is that some degree of lateral flexion and rotation can be applied to the flexion forces. Complaints expressed by the patient during manual traction can be expected if the forces, angulation, or degree is unacceptable to the patient.

The choice of continuous versus intermittent traction has not been objectively documented and currently is determined by the preference of or acceptance by the patient. Some patients feel more able to relax during intermittent traction than during continuous traction. In some patients,

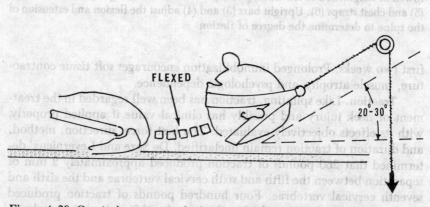

Figure 4–28. Cervical traction applied to the supine patient, causing cervical spine flexion with the angle of pull between 20 and 30 degrees.

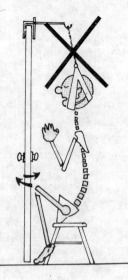

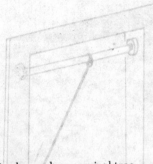

Figure 4-29. Ineffective home door cervical traction. The patient is too close to the door to receive correct neck flexion angle. The door freely opens and closes, not permitting constant traction. The patient cannot extend the legs or assume a comfortable position. This type of home traction is not recommended.

repeated tension upon the irritable muscle spindles initiates more reflex contraction (spasm).

Clinical cervical traction is useful to evaluate the patient's response and to determine the optimum amount, angle, and duration of traction. Once these measurements are established as being beneficial, home cervical traction is usually indicated.

Cervical traction equipment applied to the door often proves to be ineffectual and difficult to maintain (Fig. 4-29). A more acceptable method of home traction is placement of a chinning bar between door jams, as shown in Figure 4-30. In this type of traction, the amount and the angle of traction can be varied by the patient.

Manipulation and Mobilization. The basis for manipulation as a method of treatment remains unproven. There are numerous concepts of manipulation advocated, including:

1. "unlocking" of a facet joint. The facet allegedly becomes "locked" by:
    a. Entrapped synovial capsule.
    b. Entrapped meniscus.
    c. Assymmetric facet approximation with concomitant synovitis.
    d. "Subluxation."
2. Neurologic reflex release of neurogenic spasm via spindle effect upon the muscle.
3. Ligamentous and capsular elongation that has been contracted.

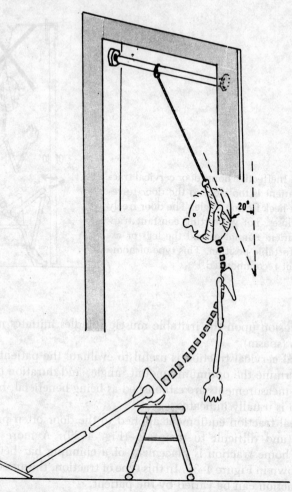

Figure 4–30. Recommended home traction from chinning bar in the sitting position.

Manipulation differs from mobilization in that during painful limitation, the joint is "forced" in a particular direction to the joint's normal range of motion. According to a different concept of manipulation, proposed by Maigne, the joint is manipulated in the opposite, unrestricted, and painless direction. This technique "unlocks" the joint by virtue of painfree contralateral movement of the joint.

Mobilization is a form of manipulation in that external force is applied to the cervical spine to regain range of motion, which allegedly realigns the spine. Mobilization differs in that no "thrust" of the joint is given upon reaching the passive range of motion.

Both manipulation and mobilization are passive procedures (i.e., done *to* the patient by an external force). Mobilization can be performed simultaneously with the patient's voluntary contraction and relaxation of the involved muscles.

Achieving segmental mobilization or manipulation of a "precise" joint is claimed by some practitioners. Movement of a functional unit such as C_5 to C_6 or C_2 to C_3 is implied. These claims are based more on achieving relief of symptoms and regaining motion than upon verification by radiologic confirmation.

Either form of passive movement can be preceded by the application of ice or heat. Any mobilization must be preceded by careful neurologic, orthopedic, and radiologic examination to rule out any complicating condition.

Physical Therapy. Modalities such as heat, ice, ultrasound radiation, and infrared radiation have been advocated in the treatment of acute neck injury. Usually, ice applied early in the acute condition relieves the pain, spasm, and inflammation. In a later stage, heat decreases the accumulation of toxic substances in the inflamed muscles; decreases the histamine, prostaglandins, substance P, and so forth; decreases the adhesions of collagen fibers of the capsules; and relieves pain.

Massage, if tolerated, produces the same results that heat does. Ultrasound radiation is essentially a form of deep heat that is valuable if used as an adjunct therapy. Unfortunately, ultrasound therapy is often used the sole modality—an approach that is bound to fail. The value of *any* modality is to increase *active movement*, which means *active exercise* to improve flexibility, strength, and posture.

Exercise. Exercises should be considered early in treating hyperflexion-hyperextension injury of the cervical spine.

Often, weakness of the neck flexors is present initially. This weakness represents a reflex inhibition due to acute spindle system activation, which prevents the patient from lifting the head from the pillow. This weakness causes difficulty in maintaining the erect neck posture and eliminates reflex inhibition of the spastic neck extensors. The sooner the flexors begin contracting, the sooner normal neuromuscular activity returns.

The exact neurologic mechanism of this flexor "paresis" remains unconfirmed, but it may be a reflex spindle reaction to overstretched muscle or to facet capsular stretching. Also, it is possibly aggravated in patients who had pre-accident weakness of the neck flexor muscles.

The basis for initiating early exercises is to maintain good tone and proprioception, prevent immobilization contracture, and restore normal range of motion.

Initially, isometric exercises, involving contraction of muscles without joint movement, are desirable and tolerated by the patient. They help the muscles regain some voluntary control, and they help to remove the accumulated nociceptive chemicals from the inflamed tissues and muscles (as noted in Chapter 2).

Isometric muscular contraction can be started within a few days of the injury, even though the patient is experiencing some pain. If injury to a joint has occurred, the isometric muscular contraction should not be detrimental. Except in cases of a torn muscle or tendon, there are very few contraindications to beginning isometric exercises early.

Isometric muscular contraction stretches the capsule to a slight degree, compresses the cartilage, strengthens the tendons and ligaments, and rids the muscles of their toxic substances. More important, *active exercise*, even though it is isometric, helps the patient become actively involved in the process of recovery.

Following the initial isometric exercises, an active assisted exercise can gradually be instituted. In this exercise, the patient voluntarily contracts the muscle isometrically with assisted gradual passive range of motion. This exercise is performed to the tolerance of the patient and to the physiologic limit of articular involvement.

Exercise can be preceded by local application of ice and can be followed by application of heat (or vice versa, as best tolerated by the patient).

The active assisted exercise should be gradually replaced by isometric active exercise. The patient determines the strength, force, and range of movement, but is guided by the therapist to minimize anxiety and "fear of damage."

Ultimately, some resistance should be applied to the muscular contraction to develop strength and endurance, achieve full range of motion, and involve the patient in recovery. Active contraction of the muscles and the simultaneous elongation of the tendons, capsules, and ligaments initiate the proprioceptive response, which enhances functional restoration.

In acute injury of the cervical spine, therapy should initially be concentrated on the neck flexors. They usually have a reflex inhibition and simultaneously inhibit reciprocal relaxation of the "spastic" antagonists. This "paresis" of the flexors may result from articular distension and injury and must be corrected early. Persistent weakness of the neck flexors has been documented in patients whose neck pain continues for months after the accident.

Isometric then isotonic exercises should concentrate on the short (occipitocervical) flexor muscles and the long (cervical) extensor muscles. Proper function of these muscles not only assures range of motion, but is vital in regaining proper posture.

Restoration of complete range of motion is ultimately desirable, but should be attempted gradually. At first, the reason for limitation is protective muscle "spasm," but persistence of limited range of motion may occur

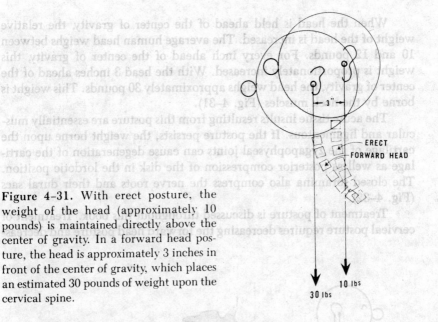

ERECT
FORWARD HEAD

10 lbs
30 lbs

Figure 4-31. With erect posture, the weight of the head (approximately 10 pounds) is maintained directly above the center of gravity. In a forward head posture, the head is approximately 3 inches in front of the center of gravity, which places an estimated 30 pounds of weight upon the cervical spine.

because of muscular contracture of connective tissue and articular contracture. Prevention of this problem is more easily attained than is gradual recovery.

Passive range of motion can be accomplished with the therapist's assistance. Gentle stretching or mobilization is aided by application of ice or a vasocoolant (the "spray and stretch" technique) preceding the stretch.

Posture. As previously mentioned, proper posture must be regained or relearned early. Imparied posture following an accident may be a protective mechanism that can rapidly become an accepted body position. Posture also represents "body language" that is unconsciously employed by the injured patient to convey to any observer the presence and degree of sustained injury.

Extension of the cervical spine has been shown to approximate the posterior elements of the spine, with narrowing of the foramina and approximation of the facet joints. Both of these changes can entrap the nerve roots and their dura or evoke pain from an inflamed subluxed facet joint. This position of cervical extension is one form of "posture," and it may result from occupational conditions.

"Forward head posture" increases the lordosis and places strain upon the erector spinae (extensor) muscles by placing the head ahead of the center of gravity. This posture may be induced occupationally, result from previous "poor" posture, or be used to convey emotions such as anger, anxiety, impatience, or depression.

When the head is held ahead of the center of gravity, the relative weight of the head is increased. The average human head weighs between 10 and 12 pounds. For every inch ahead of the center of gravity, this weight is proportionately increased. With the head 3 inches ahead of the center of gravity, the head weighs approximately 30 pounds. This weight is borne by the neck muscles (Fig. 4–31).

The acute tissue insults resulting from this posture are essentially muscular and ligamentous. If the posture persists, the weight borne upon the cartilage of the zygapophyseal joints can cause degeneration of the cartilage as well as posterior compression of the disk in the lordotic position. The closed foramina also compress the nerve roots and their dural sacs (Fig. 4–32).

Treatment of posture is discussed throughout this book. Treatment of cervical posture requires decreasing the forward head posture and decreas-

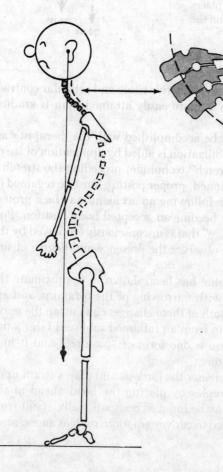

Figure 4–32. Forward head posture of depression. The depressed person has a rounded back and forward posture, which increase the cervical lordosis. This increased lordosis approximates the posterior articulations and narrows the intervertebral foramina (*upper right*).

ing the lordosis. The patient can learn to maintain a "concept of proper posture" throughout daily activities, including occupational activities, by carrying a small weight upon the head (Fig. 4–33). This exercise gives the patient the sensation of proper posture, and when the head is brought forward of the center of gravity, the weight becomes very heavy and "reminds" the person to resume the correct posture.

DEGENERATIVE JOINT DISEASE

As the uncovertebral joints undergo hypertrophy, and ultimately calcification, they tend to protrude posteriorly into the intervertebral foramen (Fig. 4–34) and thus decrease the diameter of the foramen and en-

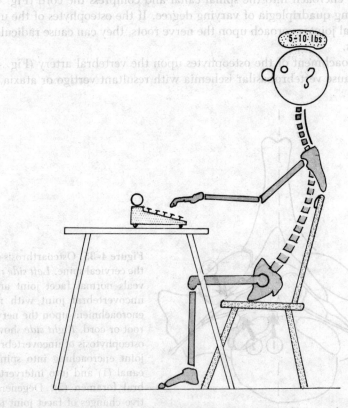

Figure 4–33. Distraction exercise for posture training. With a weight of 5 to 10 pounds within a sandbag upon the head, the erect posture is maintained, and the cervical lordosis is minimal. Proprioceptive concept of posture is learned with no effort.

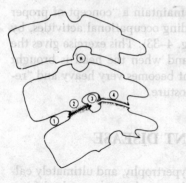

Figure 4-34. Osteophytosis of the cervical functional unit. The *top* unit (N) depicts the normal relationship. The *bottom* unit reveals a narrowed degenerated intervertebral disk (1) with hypertrophy of joints of Luschka. There are formed osteophytes protruding into the intervertebral foramen (2). The posterior facets undergo osteoarthritic changes (3) and cause posterior osteophytes to protrude into the foramen (4). Local pain and restricted motion result, as does nerve root compression within the foramen or cord compression within the canal.

croach upon the nerve roots. If the protrusion is ventral, the uncovertebral joints can encroach into the spinal canal and compress the cord (Fig. 4-35), causing quadriplegia of varying degree. If the osteophytes of the uncovertebral joints encroach upon the nerve roots, they can cause radicular symptoms.

Encroachment of the osteophytes upon the vertebral artery (Fig. 4-36) can cause vertebrobasilar ischemia with resultant vertigo or ataxia.

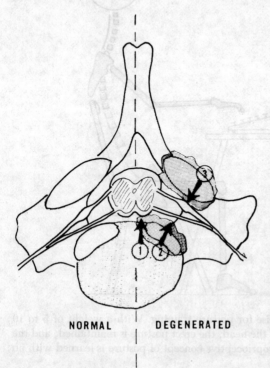

NORMAL | DEGENERATED

Figure 4-35. Osteoarthrosis of the cervical spine. *Left side* reveals normal facet joint and uncovertebral joint with no encroachment upon the nerve root or cord. *Right side* shows osteophytosis of uncovertebral joint encroaching into spinal canal (1) and into intervertebral foramen (2). Degenerative changes of facet joint (3) cause encroachment into foramen. All three osteophytes can cause neurologic deficit of cord or nerve root.

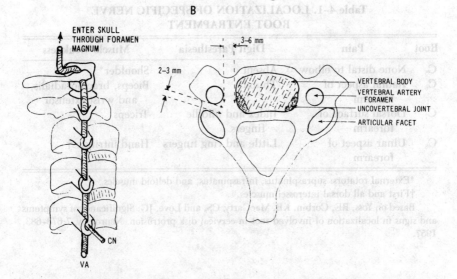

Figure 4–36. Passage of the vertebral artery through the cervical spine. *A*, Angulation at the C_1 to C_2 segment. *B*, Relationship to the uncovertebral joints and the facets.

Causation of so-called **osteoarthritis** (more properly called **osteoarthrosis**) is unclear. Compression of the cartilage has been claimed to precede osteoarthrosis, as has the slackening of the longitudinal ligaments, as the intervertebral disk dehydrates and narrows.

CERVICAL RADICULOPATHY

The so-called "diskogenic syndrome" with nerve root symptoms (radiculopathy) is very common as a complaint, but is frequently inappropriately diagnosed.

Pain or paresthesia, with or without motor weakness of a precise nerve root, may occur from nerve root entrapment within the foramen and gives subjective and/or objective dermatome or myotome distribution (Table 4–1).

Intervertebral disk herniation is considered a primary cause of nerve root entrapment, but disk herniation of either the nucleus or the annulus is rare. Nerve root entrapment may occur from:

1. Acute injury including traction or compression
2. Facet synovitis

Table 4-1. LOCALIZATION OF SPECIFIC NERVE
ROOT ENTRAPMENT

Root	Pain	Digit Paresthesia	Muscle Weakness
C_5	None distal to elbow	Absent	Shoulder*
C_6	Radial aspect of forearm	Thumb	Biceps, brachioradialis, and wrist extensor
C_7	Dorsal surface of forearm	Index and middle fingers	Triceps
C_8	Ulnar aspect of forearm	Little and ring fingers	Hand intrinsics†

*External rotators: supraspinatus, infraspinatus, and deltoid muscles.
†First and all dorsal interossei muscles.
Based on Yoss, RE, Corbin, KB, MacCarty, CS, and Love, JG: Significance of symptoms and signs in localization of involved root in cervical disk protrusion. Neurology, 7:673–683, 1957.

3. Pre-existing degenerative changes aggravated by acute injury
4. Annular bulge from torque injury.

The treatment of a neuropathy from cervical injury is the same as that for the cervical injury. Surgical intervention is considered in the following circumstances:

1. There is persistent or increasing neurologic deficit.
2. There are long tract signs.
3. There is persistent intractable pain and no benefit is obtained from appropriate conservative treatment.

No surgical intervention should be contemplated in the absence of objective neurologic signs, and in the lack of confirmation of neuropathy through significant laboratory tests, such as myelography, computerized tomography (CT) scanning, magnetic resonance imaging (MRI), and electromyelography (EMG). Any test considered pertinent must explain the symptoms claimed by the patient.

HEADACHE

Headache is usually defined as a discomfort in the forehead or scalp area, excluding the face, and below the eyebrows. The term **cephalalgia** is appropriate.

The area of "headache" is frequently that supplied by the greater and lesser occipital nerves, which are derived from the second cervical roots and the trigeminal nerves.

Headache may occur from any cranial tissue: the muscles, the scalp, the superficial blood vessels, and even intracranial contents. It is not within the province of this text to discuss all the causes of headache, but some aspects of head pain considered to originate from the cervical spine will be discussed. It has been estimated[7] that the greatest cause of disability from pain requiring treatment is head pain. A major cause of disability is low back pain, but head pain is also significant. The percentage of headaches whose cause involves the cervical spine has not been clarified, but is considered to be large.

Intracranial headache may be acute, and may be serious or even life-threatening. Extracanial headache may also be acute or chronic, mild or severe, but it is rarely, if ever, life-threatening. At worst, it is disabling and unpleasant. The majority of chronic headaches are not due to organic disease and are best resolved by appropriate tests and treatment.

Many extracranial headaches are vascular in origin. All large blood vessels are sensitive to pain produced by distension or displacement. The so-called migraine headache is such a vascular entity resulting from vasodilatation and vasoconstriction. Headaches, however, may also have a cervical basis and the diagnosis and appropriate treatment require concentration on the cervical spine.

Headache, claimed to result from cervical spondylosis, is usually occipital and unilateral. Frequently, it originates in the back of the neck and spreads to the occipital region. It may radiate to the frontal and temporal regions. This pain is usually described as aching, nagging, and wearing, rather than the throbbing or explosive, as claimed by sufferers of vascular headache. Often, posture and neck and head movement or position, influence the headache.

There may be a nerve radicular component to pain from the cervical spine or a relationship to the neck muscles that attach to the occiput. The precise mechanism of pain remains obscure, but conservative measures benefit patients with headache involving the cervical spine. The physician should remain objective regarding "what" is being accomplished in the treatment of occipital headaches that are considered to be caused by cervical spine disease. There have been several concepts regarding the mechanism of occipital neuralgia—the mechanism of headache from the cervical spine. Bogduk[8] wrote a provocative article that casts more light on this clinical entity.

Occipital neuralgia resulting from irritation of the greater occipital nerve or its roots caused by trauma to the C_2 dorsal ramus has been accepted as the nerve route of the cephalgia (Fig. 4–37). Controversy exists regarding the source and location of the irritation.

In 1949, Hunter and Mayfield[9] claimed that the posterior ramus of C_2

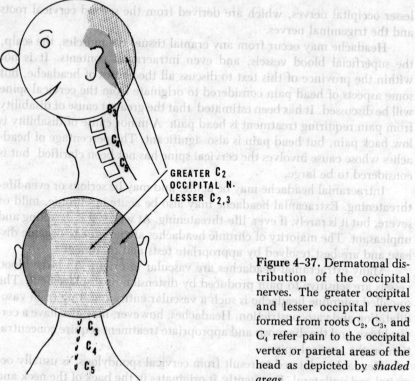

GREATER C₂
OCCIPITAL N.
LESSER C₂,₃

Figure 4–37. Dermatomal distribution of the occipital nerves. The greater occipital and lesser occipital nerves formed from roots C_2, C_3, and C_4 refer pain to the occipital vertex or parietal areas of the head as depicted by *shaded areas.*

was crushed between the posterior arch of the atlas and the lamina of the axis as a result of excessive extension of the head with simultaneous rotation toward that side. Bogduk[8] showed clearly that the nerve cannot be anatomically mechanically compressed from this maneuver. The concept that the nerve (C_2) emerged through the posterior atlanto-axial membrane was also refuted, as the nerve has lateral access to that membrane (Fig. 4–38).

The concept of compression of the greater occipital nerve at its exit point under the mastoid bone, between the fibers of the sternocleidomastoid and the upper trapezius, was also refuted, as any contraction of these muscles from either "tension" or contraction draws the sling (considered to be the band compressing the nerve) downward and "away" from the nerve.

According to Bogduk,[8] the mechanism of occipital neuralgia is as follows. The C_2 ganglion lays under the posterior arch of the atlas, medial to the articular capsule of the lateral bodies of the atlas and the axis. The nerve then branches into the dorsal and the ventral rami, which supply

Figure 4-38. Greater superior occipital nerve (C_2). The extradural C_2 nerve roots emerge from the cervical spine as shown on *right*.

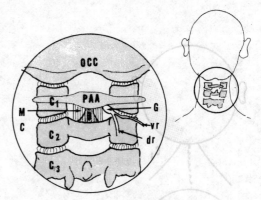

C₁— atlas
C₂— axis
C₃— third cervical vertebra
OCC— occiput
M— posterior atlanto-axial membrane
O— odontoid of axis
PAA— posterior arch of atlas (C₁)
G— ganglion of C_2 nerve
C— capsules of cervical joints
vr— ventral root of C_2
dr— dorsal root of C_2

sensory fibers to the capsule of that joint. Probably the occipital neuralgia, especially that following rear-end automobile collision, in which the upper cervical segment is violently flexed, extended, and rotated, results from a "subluxation" of that joint with excessive capsular stretch. The resultant occipital cephalalgia probably represents referred pain from irritation of the dorsal root of the greater occipital nerve as it penetrates the joint capsule.

Treatment requires early immobilization for a few days with local heat or ultrasound radiation (for heat penetration), then traction to ensure separation of the upper cervical vertebrae. Injection of the occipital nerve at its point of emergence at the occipital groove relieves symptoms, but does not alter their cause, which is articular.

Headache considered to originate from the lower cervical segments (C_4 through C_7) probably originates from irritation of the posterior primary divisions, which transmit sensations to the trigeminal nucleus in the cord.

Headaches can occur from irritation of the cervical extensor muscles as well as from tension of the occipital muscles. Research has revealed that injection of an irritating substance into the erector spinae muscles at specific segmental levels causes specific headache patterns (Fig. 4-39).

"Trigger" areas, zones of muscular nodularity and exquisite tenderness when pressed, are found in the sternocleidomastoid, splenius capitis,

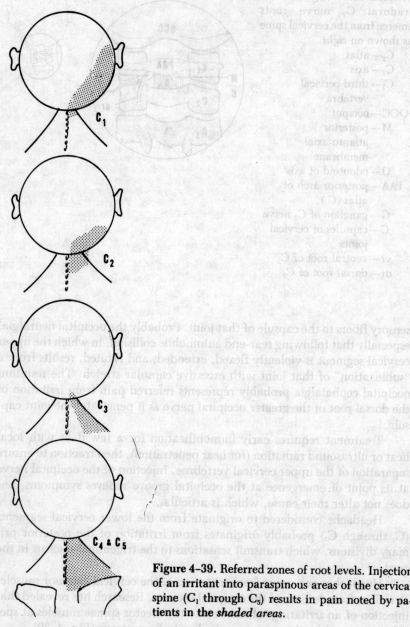

Figure 4–39. Referred zones of root levels. Injection of an irritant into paraspinous areas of the cervical spine (C_1 through C_5) results in pain noted by patients in the *shaded areas*.

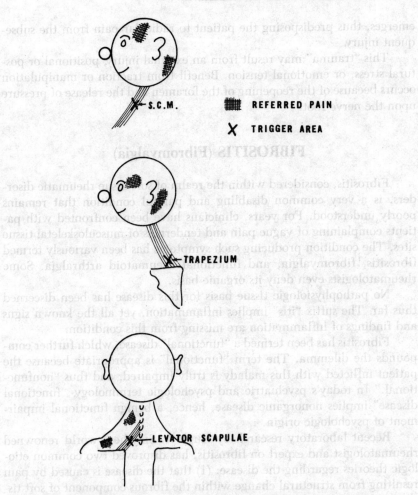

Figure 4-40. Trigger points and referred pain. Tender areas in various areas of the neck and shoulder, when irritated, can refer pain to distal sites.

temporalis, masseter, and trapezius muscles, which "refer" pain to occipital areas (Fig. 4-40). Injection of an anesthetic substance into these trigger areas obliterates the local pain and tenderness as well as pain in the referred area. These painful trigger areas are known as "myofascial" pain triggers.

Headaches originating from osteophytes in the cervical spine may occur, but the pain occurs from injury or stress of a cervical spine that already has osteophytes. The presence of osteophytes does not cause the head pain; their presence narrows the foramen through which the nerve root emerges, thus predisposing the patient to radicular pain from the subsequent injury.

This "trauma" may result from an external injury, positional or postural stress, or emotional tension. Benefit from traction or manipulation occurs because of the reopening of the foramen and the release of pressure upon the nerve root.

FIBROSITIS (Fibromyalgia)

Fibrositis, considered within the realm of soft tissue rheumatic disorders, is a very common disabling and painful condition that remains poorly understood. For years, clinicians have been confronted with patients complaining of vague pain and tenderness of musculoskeletal tissue sites. The condition producing such symptoms has been variously termed fibrositis, fibromyalgia, and functional rheumatoid arthralgia. Some rheumatologists even deny its organic basis.

No pathophysiologic tissue basis for this disease has been discerned thus far. The suffix "itis" implies inflammation, yet all the known signs and findings of inflammation are missing from this condition.

Fibrositis has been termed a "functional" disease, which further compounds the dilemma. The term "functional" is appropriate because the patient inflicted with this malady is truly impaired, and thus "nonfunctional." In today's psychiatric and psychologic terminology, "functional disease" implies nonorganic disease, hence, a benign functional impairment of psychologic origin.

Recent laboratory research by Hugh A. Smythe,[10] world renowned rheumatologist and expert on fibrositis, has disproved two common etiologic theories regarding the disease: (1) that the disease is caused by pain resulting from structural change within the fibrous component of soft tissue, and (2) that the disease is caused by pain from sustained muscle tension. Smythe suggests that the symptoms of fibrositis may represent not a rheumatic disease, but a disturbance of pain modulation within the central nervous system pathways.

The basis for his assumption is the frequency of deep sleep disturbances in patients having this syndrome. The experimental deprivation of delta (deep sleep) in normal subjects has resulted in symptoms of fibrositis.

Clinically, the patient presents with symptoms of diffuse aching, tenderness, and stiffness. This aching and stiffness, noted more frequently in the morning, are associated with headaches, poor rest patterns from sleep (i.e., patient awakes "tired" after an apparently adequate night's sleep), exhaustion, and fatigue. These symptoms appear to be adversely influenced by inclement weather changes.

The symptoms are usually bilateral and are noted in the region of the shoulder (upper trapezius region), the anterior midline region of the neck behind the sternocleidomastoid muscle, the region of the lateral epicondyle of the upper forearm ("tennis elbow" site), the medial knee joint region, and the posterior pelvis in the vicinity of the iliac crest. Symptoms in other sites are noted, but those sites mentioned appear to be the most frequent. The syndrome of temporomandibular joint (TMJ) arthralgia has recently been noted to occur in association with the syndrome of fibrositis.

A personality characteristic in patients with fibrositis has been identified. These patients tend to be self-driven, intense, compulsive, and fervently involved in numerous activities. Their personalities strongly resemble those of "migraine" patients. They are also very sensitive and responsive to external stimuli such as cold, heat, humidity, and noise.

Psychiatric studies have attempted to associate this syndrome with emotional depression that is overt, latent, or masked. Such association has been, in part, refuted, as these fibrositic patients are not withdrawn, but rather actively involved in numerous activities. After long periods of frustration from treatment failures and persistent pain, some patients exhibit depression.

Even though the concept that "tension myalgia" is the major cause of fibrositis has been refuted, some clinicians are persuaded that emotional tension contributes to conditions resembling fibrositis. This tension myositis is also markedly influenced by postural stresses, which in turn are influenced by occupational and emotional stresses.

Quantitative electromyographic investigation causing muscle pain from induced "spasms" has been well documented by deVries.[11] He induced muscle strain resulting in spasm and claimed that relief occurred, as evidenced clinically and by EMG, after the application of ice and gradual sustained stretching. The theories of Travell[12] and Simon[12] add support to this concept.

Diagnosis

The diagnosis of fibrositis is clinical. Obtaining a proper history is mandatory, and appropriate examination is required to identify the "tender" spots. Increased sensitivity is noted over the ligaments, tendons, and periarticular tissues. This deep tenderness is pathognomonic of fibrositis and helps refute the concept that muscle is the exclusive site of tenderness.

Laboratory confirmation is absent. Blood counts and rheumatoid factors are negative. An elevated erythrocyte sedimentation rate (ESR)—a Westergren ESR greater than 50 in 1 hour—implies polymyalgia rheumatica (PMR). Biopsies of the tender nodules have been nondiagnostic. Referred pain with dermatomal nerve root radiation presents a challenge in differential diagnosis. These referral patterns have been amply discussed within this chapter.

Treatment

Treatment of fibrositis taxes the patience and expertise of the physician and may be frustrating to the patient. A good rapport between patient and physician is mandatory, as this condition is refractory to treatment of the usual type.

Fibrositis is chronic and rarely abates spontaneously or completely. The physician must emphasize to the patient, however, that fibrositis is neither serious nor a potential "crippler." This point may offer little solace to the suffering patient for whom the main objective is to achieve comfort during daily activities.

Stress management is probably the most beneficial approach to treating this condition. Life stresses must be recognized, and "coping" mechanisms developed. Lifestyle must be altered. A meaningful but realistic schedule of activities must be established.

The detrimental sleep pattern abnormality must be altered. Tricyclic drugs taken at bedtime may be valuable if their sedative effect does not continue during the waking hours. Benzodiazepines are short-acting and may replace the tricyclics. The physician must ensure that use of addictive drugs is not implemented or maintained.

Beneficial effects of local treatment of the tender site are limited, although desirable. Techniques using ice, passive and active stretching, deep massage, neuroprobe, acupuncture, and so forth are of temporary value and may be used judiciously. Addiction to these modalities may be induced, to the consternation of the health policy financial carriers, and even to the patients themselves.

An active exercise program is of great value, but patients with fibrositis have been known to be remiss in adhering to such programs. Correct modification of posture in everyday activities, and in professional pursuits, is mandatory and beneficial, as are regaining and maintaining general tissue flexibility.

Stress management, hypnosis, progressive relaxation, psychotherapy, and counseling are of value. Biofeedback also assists in relieving not only the cause but also the effect of the muscular tension. Biofeedback may employ sounds, feelings, bells, lights, and so forth in an attempt to help the patient gain control of the effects of the autonomic and somatic nervous systems on the musculoskeletal system.

All injuries or insults to the cervical spine cause some reactive "protective" muscular spasm of the erector spinae muscles. Nociceptive substances are released as a result of this muscular tension, and ultimately, the spasm, originally intended to splint the injured part, becomes the source of pain.

REFERENCES

1. Cailliet, R: Neck and Arm Pain, ed 2. FA Davis, Philadelphia, 1981.
2. Cloward, RB: The clinical significance of the sinu-vertebral nerve of the cervical spine in relation to the cervical disc syndrome. J Neurol Neurosurg Psychiatry 23:321, 1960.
3. Holt, EP: Fallacy of cervical discography. JAMA 188:799, 1964.
4. De Seze, S, and Levernieux, J: Les tractions vertebrales; priemeres etudes experimentales et resultats therapeutiques d'apres une experience de quatres annes. Semaines des Hopitaux (de Paris) 27:2085–2104, 1951.
5. Crue, BJ: Importance of flexion in cervical traction for radiculitis. USAF Med J 8:375–380, 1957.
6. Colachis, SC, and Strohm, BR: Cervical traction: Relationship of traction time to varied tractive force with constant angle of pull. Arch Phys Med Rehabil 46:815–819, 1965.
7. Nuprin Report of Bristol Myers Pharmaceutical Company.
8. Bogduk, N: Headaches and cervical manipulation. Med J Aust, July 28, 1979, pp 65–66.
9. Hunter, CR, and Mayfield, FH: Role of the upper cervical roots in the production of pain in the head. Am J Surg 78:743–749, 1949
10. Smythe, HA: Fibrosis and the Referred Pain Syndromes. Rheumatology Forum, vol 3, no. 1. Pfizer, New York, 1985.
11. deVries, HA: Quantitative electromyographic investigation of the spasm theory of muscle pain. Am J Phys Med, 45:119–143, 1966.
12. Travell, JG and Simons, DG: Myofascial Pain and Dysfunction: The Trigger Point Manual. Williams & Wilkins, Baltimore, 1983.

BIBLIOGRAPHY

Birk, L (ed): Biofeedback: Behavior Medicine. Grune & Stratton, New York, 1973.
Blumenthal, L: Injury to the cervical spine as a cause of headache. Postgrad Med 56:147–153, 1974.
Bogduk, N: Headaches and cervical manipulation. Med J Aust, July 28, 1979, pp 65–66.
Braaf, MM, and Rosner, S: Trauma of the cervical spine as cause of chronic headache. J Trauma 15:441–446, 1975.
Brain, L: Some unsolved problems of cervical spondylosis. Br Med J 1:771–776, 1963.
Caldwell, JW, and Krusen, EM: Effectiveness of cervical traction in treatment of neck problems: Evaluation of various methods. Arch Phys Med Rehabil 43:214–222, 1962.
Campbell, SM, Clark, S, Tindall, EA, et al: Clinical characteristics of fibrositis. Arthritis Rheum 26:817–824, 1983.
Cloward, RB: The clinical significance of the sinu-vertebral nerve of the cervical spine in relation to the cervical disc syndrome. J Neurol Neurosurg Psychiatry 23:321, 1960.
Colachis, SC, Jr, and Strohm, BR: Effect of duration of intermittent cervical traction on vertebral separation. Arch Phys Med Rehabil 47:353, 1966.
Coppola, AR: Disease of the cervical spine and nerve root pain. Va Med Mon 101:199–201, 1974. (A concise report on the symptoms, diagnosis, and treatment of lesions of the cervical spine.)

DePalma, AF, et al: Study of the cervical syndrome. Clin Orthop 38:135–142, 1965. (A comparison of the results of conservative and surgical treatment for relief of cervical syndromes leads these authors to feel that surgical intervention is justified more often than is generally believed.)

deVries, HA: Quantitative electromyographic investigation of the spasm theory of muscle pain. Am J Phys Med 45:119–134, 1966.

Du Toit, GT: The post-traumatic painful neck. Forensic Sci 3:1–18, 1974. (The common mechanisms of injury, diskography, arteriography, neurologic signs, therapy, fusion, and radiologic investigation are discussed.)

Fielding, JW, Burstein, AH, and Frankel, VH: The nuchal ligament. Spine 1:3–14, 1976.

Frykholm, R, et al: On pain sensations produced by stimulation of ventral roots in man. Acta Physiol Scand 29:455, 1953.

Gukelberger, M: The uncomplicated post-traumatic cervical syndrome. Scand J Rehabil Med 4:140–153, 1972. (Types of trauma, symptoms, and functional radiologic examination are described, and different types of radiologic changes are reported and illustrated.)

Hartman, JT: A conversation with JT Hartman: The cervical orthosis—does it immobilize? Orthop Review 5:53–57, Oct 1976.

Holt, EP: Fallacy of cervical discography. JAMA 188:799, 1964.

Hunter, CR, and Mayfield, FH: Role of the upper cervical roots in the production of pain in the head. Am J Surg 78: 743–749, 1949.

Jackson, R: The Cervical Syndrome, ed 4. Charles C Thomas, Springfield, IL 1971.

Jacobson, E: Progressive Relaxation, ed 2. University of Chicago Press, Chicago, 1938.

Krout, RM, and Anderson, TP: Role of anterior cervical muscles in production of neck pain. Arch Phys Med Rehabil 47:603–611, 1966.

Lalli, JJ: Cervical vertebral syndromes. J Am Osteopath Assoc 72:121–128, 1972. (A review of the current medical literature on cervical vertebral syndromes and their treatment is presented along with a useful discussion of the functional anatomy, symptoms, diagnosis, and conservative medical management.)

Lance, JW: Headache. In Critchley, M (ed): Scientific Foundations of Neurology. FA Davis, Philadelphia, 1972, pp 169–175.

Lourie, H, et al: The syndrome of central cervical soft disk herniation. JAMA 226:302–305, 1973. (Case histories illustrating the syndrome of central cervical soft disk herniation are presented, and attention is called to possible pitfalls in diagnosis and management.)

Macdonald, AJR: Abnormally tender muscle regions and associated painful movements. Pain 8:197–205, 1980.

MacNab, I: The whiplash syndrome. Clin Neurosurg 20:232–241, 1973. (Doubts about the validity of a diagnosis of "litigation neurosis" precede a discussion of the mechanics of the cervical spine.)

MacNab, I: Acceleration extension injuries in the cervical spine. In American Academy of Orthopedic Surgeons Symposium of the Spine. CV Mosby, St. Louis, 1969, pp 10–17.

Maigne, R: Orthopedic Medicine. Charles C Thomas, Springfield, IL, 1972.

McCall, IW, et al: The radiologic demonstration of acute lower cervical injury. Clin Radiol 24:235–240, 1973. (This paper emphasizes the limitations of anteroposterior and lateral radiographs and describes a simple technique for supine oblique examination of the immobilized patient.)

Miglietta, O: Action of cold on spasticity. Am J Phys Med 52:198–204, 1973.

Miles, WA: Discogenic and osteoarthritic disease of the cervical spine. J Natl Med Assoc 66:300–304, 1974. (A brief review of the anatomy and physiology of the cervical spine is followed by discussion of symptoms, radiographic and pathologic findings, and radiographic assessment of operative intervention.)

Nicholas, GG: Initial management of injury to the neck. Pa Med 77:47–48, 1974. (Emergency measures are described to provide an adequate airway and to control hemorrhage in patients with neck injury.)

Ostfeld, AM: The Common Headache Syndromes: Biochemistry Pathophysiology Therapy. Charles C Thomas, Springfield, IL, 1962.

Rothman, RH, and Simeone, FA (eds): The Spine, Vols 1 and 2, WB Saunders, Philadelphia, 1975.

Sano, K, et al: Correlative studies of dynamics and pathology in whip-lash and head injuries. Scand J Rehabil Med 4:47–54, 1972.

Severy, DM, Mathewson, JH, and Bechtol, CO: Controlled automobile rearend collision: An investigation of related engineering and medical phenomena. Can Serv Med J 11:727, 1955.

Stainsbury, P, and Gibson, JE: Symptoms of anxiety and tension and accompanying physiological changes in the muscular system. J Neurol Neurosurg Psychiatry 17:216–224, 1954.

Veleanu, C, and Diaconescu, N: Contribution to the clinical anatomy of the vertebral column consideration on the stability and the instability at the height of the "vertebral unit." Anat Anz 137S:287–295, 1975.

Weinberger, LM: Traumatic fibromyositis: A critical review of an enigmatic concept. West J Med 127:99–103, Aug 1977.

Wiberg, G: Back pain in relation to the nerve supply of the intervertebral disc. Acta Orthop Scand 19:211, 1949.

CHAPTER 5
Neurovascular Compression Syndromes

The symptoms of **thoracic outlet syndrome**, a poorly understood syndrome, are pain, numbness, and tingling in the upper extremity. The cause of the syndrome has been considered to be a compression of the neurovascular bundle (the subclavian artery and vein and the brachial plexus) between the first rib, the clavicle, and the scalene muscles. This opening is termed the **cervicodorsal outlet** or the **thoracic outlet**.

Thoracic outlet syndrome has been given many other names, including the scalene anticus syndrome, the cervicodorsal outlet syndrome, the hyperabduction syndrome, the cervical rib syndrome, the clavicocostal syndrome, and the pectoralis minor syndrome.

The cervical nerve roots of the brachial plexus descend from the cervical vertebrae and merge into trunks and cords, and ultimately form peripheral nerves (Fig. 5–1). The peripheral nerves are mostly the radial, median, ulnar, and musculoskeletal nerves of the upper extremity.

The symptoms of the syndrome have been attributed to vascular and/or neurologic compression. The latter form of compression results primarily from entrapment of the median trunks involving the eighth cervical and first thoracic roots. The symptoms are referred into the dermatomal distribution of these nerves roots.

The subclavian artery arches over the first rib behind the anterior scalene muscle and in front of the middle scalene muscle (Fig. 5–2). It proceeds under the clavicle and enters the axilla beneath the pectoralis minor muscle. The subclavian vein courses adjacent to the artery except that it passes anterior to the anterior scalene muscle and is not usually

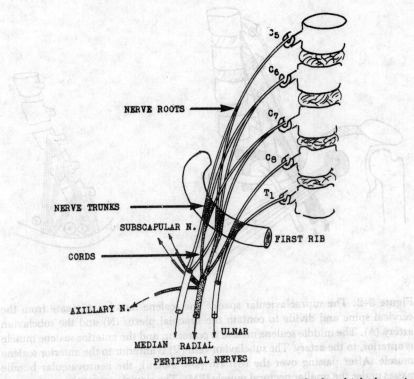

Figure 5–1. Schematic representation of brachial plexus. The brachial plexus is composed of the anterior primary rami of segments C_5, C_6, C_7, C_8, and T_1. The roots emerge from the intervertebral foramina through the scalene muscles. The roots merge into three trunks in the region of the first rib. The trunks via divisions become cords that divide into the peripheral nerves of the upper extremities.

within the foramen of the outlet. The entire neurovascular bundle is confined within a space formed by a dense fascia.

During arm and shoulder movement, the scapula moves to 45 degrees of abduction, and the glenohumeral joint permits full overhead elevation of the arm. In this movement, the axillary artery is bent 180 degrees from its vertical dependent position and is pulled across the coracoid process and the head of the humerus.

SYMPTOMS

The symptoms of a patient with a neurovascular compression syndrome depend on whether the nerve, the blood vessels, or both are compressed at the thoracic outlet. Sensory symptoms of nerve compression are

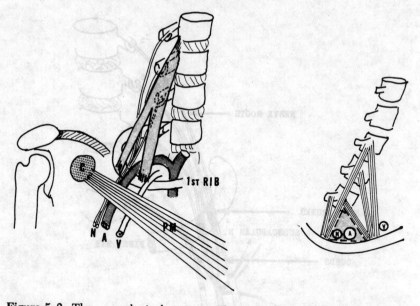

Figure 5–2. The supraclavicular space. The scalene muscles originate from the cervical spine and divide to contain the brachial plexus (N) and the subclavian artery (A). The middle scalene muscle is posterior, and the anterior scalene muscle is anterior, to the artery. The subclavian vein (V) is anterior to the anterior scalene muscle. After passing over the first rib *(not shown)*, the neurovascular bundle passes under the smaller pectoral muscle (PM). The clavicle covers the neurovascular bundle and lays parallel to the first rib. The coracoid process is labeled C.

most frequent (95%) with motor weakness less noted (10%). The symptoms are essentially pain and paresthesia. The pain and the paresthesia are usually insidious in their onset and are located in the neck, shoulder, arm, hand, and fingers. The ulnar nerve distribution is the site of complaint in 90 percent of patients (Fig. 5–3).

Symptoms attributable to vascular compression include coldness, numbness, fatigability, and some pain. Venous compression, although rare, elicits edema and a bluish discoloration. Objective findings are most frequent in vascular occlusion and often absent in neurologic compression. Loss or diminution of the radial pulse during the Adson maneuver or the hyperextended position is frequently noted. The loss of radial pulse during these maneuvers occurs often in asymptomatic individuals, so interpretation of this finding must be carefully correlated with symptoms in the history. The objective signs of nerve compression are hyperesthesia or hypalgesia of a dermatomal distribution, muscular weakness, and atrophy of a myotomal distribution.

Confirmatory diagnostic procedures now exist. Electromyographic (EMG) conduction studies of the ulnar, median, radial, and musculoskele-

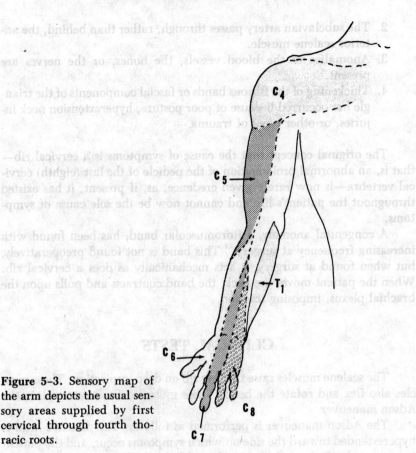

Figure 5-3. Sensory map of the arm depicts the usual sensory areas supplied by first cervical through fourth thoracic roots.

tal nerves above the clavicle can be measured and compared with measurement distal to the clavicle. A delay of conduction at the clavicle, especially if the delay is prolonged when the arm is abducted and elevated, is a confirming sign.

CAUSES

The triangle formed by the anterior scalene muscles, the middle scalene muscle, and the first rib, through which traverse the subclavian vessels and the brachial plexus, remains essentially constant in everyday activities. Symptoms of occlusion can occur as a result of various abnormalities, including the following:

1. Accessory cervical rib, with or without fibrous extension, narrows the interscalene triangle.

2. The subclavian artery passes through, rather than behind, the anterior scalene muscle.
3. Anomalies of the blood vessels, the bones, or the nerves are present.
4. Thickening of the fibrous bands or fascial components of the triangle has occurred because of poor posture, hyperextension neck injuries, or other types of trauma.

The original concept that the cause of symptoms is a cervical rib—that is, an abnormal prolongation of the pedicle of the last (eighth) cervical vertebra—is now rarely given credence, as, if present, it has existed throughout the patient's life and cannot now be the sole cause of symptoms.

A congenital anomaly, a fibromuscular band, has been found with increasing frequency at surgery.[1,2] This band is not found preoperatively, but when found at surgery, it acts mechanically as does a cervical rib. When the patient moves the neck, the band contracts and pulls upon the brachial plexus, imposing traction.

CLINICAL TESTS

The scalene muscles raise the first rib on deep inspiration. These muscles also flex and rotate the head. These muscular functions explain the Adson maneuver.

The Adson maneuver is performed as follows: The patient's head is hyperextended toward the side on which symptoms occur, and the patient is instructed to take a deep breath and hold it. The arm is held at the patient's side, and the radial pulse is palpated. A positive result is indicated when the pulse is obliterated and *the symptoms are produced*.

The costoclavicular test is performed to determine if the neurovascular bundle is compressed between the rib and the collar bone. The patient is placed in an exaggerated military posture, with the shoulders held back and down. The examiner requests the patient to take a simultaneous deep breath. The examiner may also apply a downward pressure on the patient's shoulders.

The hyperabduction syndrome is confirmed clinically by placing the patient's arms in the abducted position and pulling them backward. This movement allegedly places traction upon the pectoralis minor muscle, which causes neurovascular compression.

A "three-minute elevation test" has been advocated, and positive results have proved to be diagnostic. The patient sits erect, elevates both arms to right-angle abduction, and externally rotates the arms, placing the hands toward the ceiling and the elbows behind the body. In this position, the patient is told to open and close the fingers of both hands for 3 minutes. If the symptoms occur and the radial pulse remains palpable, neurologic entrapment is considered to occur.

Another diagnostic sign of neurologic entrapment consists of tenderness and reproduction of symptoms elicited when the examiner's thumb is deeply pressed at the supraclavicular area and lateral aspect of the cervical spine, the region of the brachial plexus.

ANCILLARY TESTS

Since it has been traditionally established that the symptoms are essentially neurologic, the use of invasive vascular tests, such as ateriography, are rarely needed. Nerve conduction studies and needle electromyography have been widely heralded but the results remain controversial. Ulnar nerve conduction velocity[3] has been largely used to confirm the indication for surgical intervention but has been considered unnecessary and ineffective by some authorities.[1,4-6] Most recently, somatosensory evoked potentials (SEP) have been suggested as being more accurate.

TREATMENT

Conservative and extensive treatment should be administered before such drastic procedures as venography, arteriography, or surgical exploration are undertaken.

Improvement of posture, stress management, change of faulty postural habits (especially occupational postural habits), and strengthening of scapular muscles must be initiated.

The fascia of the cervicodorsal (thoracic) outlet, through which all the neurovascular bundle components pass, may be thickened and contracted. These conditions may immobilize the scalene muscles, affecting the excursion of the first two ribs. These muscles must be mobilized, that is, elongated. This elongation can be accomplished with cervical traction, preferably with initial manual traction performed by a therapist followed by a home excerise program.

Manual "stretching" requires the therapist to fix the clavicle with one hand and then to pull the head slowly and progressively to one side while gradually extending both the neck and head. Force or a "bouncing" stretch is to be avoided. Stretching can be preceded by application of ice or a vasocoolant spray to the neck and supraclavicular tissues.

The treatment program for cervical diskogenic disease outlined in Chapter 4, which includes improvement of posture, may be incorporated, as indicated, in the treatment of thoracic outlet syndrome. Loss of muscle tone, strength, and endurance of the scapular elevators (trapezius muscles) has been considered to contribute to production of thoracic outlet syndrome. The syndrome is most often evident in middle-aged postmenopausal females, but may also be noted in depressed individuals, as evidenced by chronic postural defects. The syndrome may develop regardless of the cause of the depression.

The round-shouldered posture predisposes an individual to development of thoracic outlet syndrome. This posture causes gradual adaptive shortening of the flexors—the pectorals and neck flexors—and disuse atrophy of the scapular elevators and adductors. The scalene muscles also undergo adaptive shortening.

Resistive exercises to the scapular elevators (Figs. 5–4 and 5–5) must be performed for a significant period of time to regain strength and endurance of the scapular elevators. With weights held in both hands, arms are slowly, gradually, and repeatedly lifted until a position of full shoulder elevation is attained; the arms are then held in that position. During this

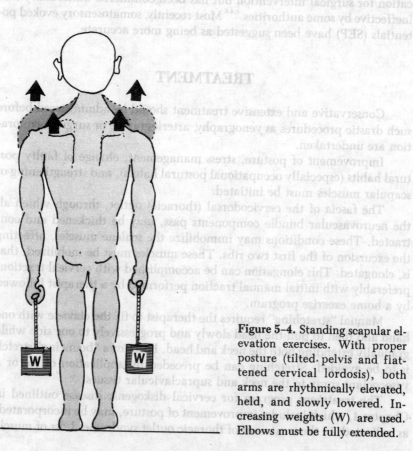

Figure 5–4. Standing scapular elevation exercises. With proper posture (tilted. pelvis and flattened cervical lordosis), both arms are rhythmically elevated, held, and slowly lowered. Increasing weights (W) are used. Elbows must be fully extended.

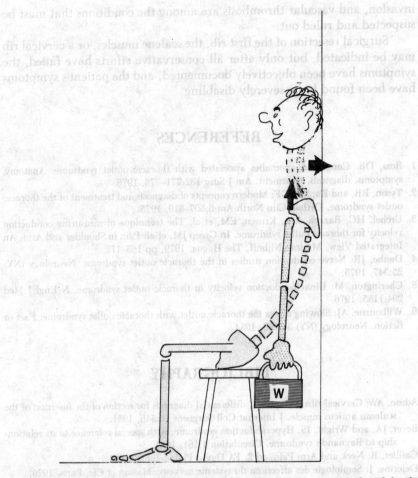

Figure 5-5. Exercise for posture and scapular elevation. Patient is seated with back to wall, and head and neck pressed against the wall, which decreases the cervical lordosis. With arms fully extended and dependent, weights (W) are lifted in a shrugging motion. Weights vary from 5 to 30 pounds.

exercise, the "forward head posture" must be avoided to maintain minimal dorsal kyphosis.

Everyday activities must be evaluated to correct faulty posture. Combating depression pharmacologically and with counseling should be an early consideration, as should stress management for extremely tense patients.

When thoracic outlet syndrome is structurally caused, as in calloused clavicular fracture or malalignment of a fractured clavicle, surgical repair or correction may be needed. Post-irradiation fibrosis, metastatic plexus invasion, and vascular thrombosis are among the conditions that must be suspected and ruled out.

Surgical resection of the first rib, the scalene muscles, or a cervical rib may be indicated, but only after all conservative efforts have failed, the symptoms have been objectively documented, and the patient's symptoms have been found to be severely disabling.

REFERENCES

1. Ross, DB: Congenital anomalies associated with thoracic outlet syndrome: Anatomy, symptoms, diagnosis, treatment. Am J Surg 132:771–778, 1976.
2. Tyson, RR, and Kaplan, GF: Modern concepts of diagnosis and treatment of the thoracic outlet syndrome. Orthop Clin North Am 6:507–519, 1975.
3. Urchel, HC, Razzuk, MA, Krusen, EM, et al: The technique of measuring conduction velocity for thoracic outlet syndrome. In Greep JM, et al: Pain in Shoulder and Arm: An Integrated View. Matinus Nijhoff, The Hague, 1979, pp 165–172.
4. Daube, JR: Nerve conduction studies in the thoracic outlet syndrome. Neurology (NY) 25:347, 1975.
5. Cherington, M: Ulnar conduction velocity in thoracic outlet syndrome. N Engl J Med 294:1185, 1976.
6. Wilbourne, AJ: Slowing across the thoracic outlet with thoracic outlet syndrome: Fact or fiction. Neurology (NY) 34:143, 1984.

BIBLIOGRAPHY

Adson, AW: Cervical ribs: Symptoms, differential diagnosis for section of the insertion of the scalenus anticus muscle. J Internat Coll Surgeons 16:546, 1951.
Beyer, JA, and Wright, IS: Hyperabduction syndrome, with special reference to its relationship to Raynaud's syndrome. Circulation 4:161, 1951.
Cailliet, R: Neck and Arm Pain, ed 2. FA Davis, Philadelphia, 1981.
Dejerine, J: Semiologie des affections du systeme nerveux. Masson et Cie, Paris, 1926.
Edgar, MA, and Nundy, S: Innervation of the spinal dura mater. J Neurol Neurosurg Psychiatry 29:530, 1966.
Lord, JW, and Rosati, LM: Neurovascular compression syndromes of the upper extremity. Clin Symp 10:1958.
Naffziger, HC, and Grant, WT: Neuritis of the brachial plexus mechanical in origin: The scalenus syndrome. Surg-Gynecol Obstet 67:722, 1938.
Nelson, PA: Treatment of patients with cervico-dorsal outlet syndrome. JAMA 27:1575, 1957.
Telford, ED, and Mottershead, S: Pressure at the cervicobrachial junction. J Bone Joint Surg 30[Br]:249, 1948.
Urschel, HC, and Razzuk, MA: Management of the thoracic-outlet syndrome: Current concepts. N Engl J Med 286:1140–1144, 1972.
Urschel, HC, Wood, RE, and Paulson, DL: Objective diagnosis (ulnar nerve conduction velocity) and current therapy of the thoracic outlet syndrome. Ann Thorac Surg 12:608–620, 1971.
Woods, WW: Personal experience with surgical treatment of 250 cases of cervicobrachial neurovascular compression syndrome. J Internat Coll Surgeons 44:273–283, 1965.

CHAPTER 6

Shoulder Pain

The shoulder contains seven joints that together may be termed the **shoulder complex.** Each joint plays a specific role in every upper extremity function, and each is a potential site of nociceptive (painful) stimuli when malfunction occurs.

Each joint will be discussed and will be related to its role in total shoulder function, malfunction, and resultant pain. The tissue site and mechanism of pain will be elucidated.

Of the seven joints in the complex, the glenohumeral joint is considered in many respects to be "the shoulder joint" and is the major site of pain and disability. It will be given major consideration.

FUNCTIONAL ANATOMY

The **glenohumeral joint** is the articulation between the rounded head of the humerus and the concave surface of the glenoid fossa. This joint can be considered to be incongruent. **Congruency** of a joint pertains to the relationship of the opposing articulating surfaces and their curvatures.

A true congruent joint would have the convexity and the opposing concavity forming perfect curves, with each point along the arc being equally opposed and equidistant to the other surface (point A in Fig. 6–1). No anatomic or engineered joint can be a completely congruent joint, as lubrication of such a joint would not be possible. The convex surface would squeeze out all the intervening fluid, and the joint would "lock" or "freeze."

179

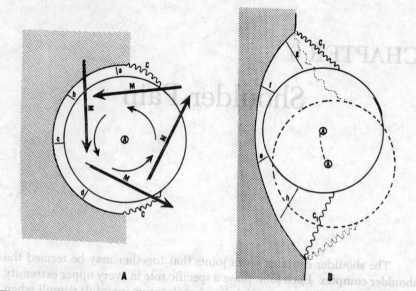

Figure 6–1. Congruous and incongruous joints. *A*, In a congruous joint, the concave and convex surfaces are symmetric. The articular surfaces are equidistant from each other at all points along their circumference (a = b = c = d, and so forth). In rotation, movement occurs about a fixed axis (A). Muscular action (M) is that of symmetric movement about this fixed axis and is needed for motion, not stability. The depth of the concave surface gives the joint stability. The capsule (C) has symmetric elongation. *B*, Incongruous joints have asymmetric articulatory surfaces. The concave surface is elongated, and the convex is more circular; thus, the distance between them varies at each point (g>f>e<h). As the joint moves, the axis of rotation (A) shifts, and joint movement is that of gliding rather than rolling. Therefore, muscles must slide the joint and simultaneously maintain stability. The capsule varies in its elongation at all levels of movement. The glenohumeral joint is an incongruous joint.

Of all the body joints, the hip joint—where the femur meets the acetabulum—is closest to being congruent. The ball-shaped (convex) femoral head fits into the ball-shaped acetabulum with only slight asymmetry, which makes the joint deep-seated and thus stable. Movement occurs about a fixed axis of rotation (point A in Fig. 6–1). The joint capsule is thus also equally distributed and restricts movement equally in all directions. Simple muscular action about the fixed axis causes flexion, extension, limited rotation, and slight abduction and adduction. The muscles need merely to mobilize the joint, but not to stabilize it.

An incongruous joint lacks these characteristics. Because of asymmetry (incongruity), the axis of rotation changes with any movement. Rather than merely moving about a fixed axis of rotation, the convex bone glides,

because the axis moves with all movements (Fig. 6–1B). The muscles must give the joint stability as well as mobility, owing to the shallowness of the "cup."

An incongruous joint affords greater mobility in exchange for stability. Muscular action is therefore also more complex, as it must provide stability, and move the joint in numerous directions. It must allow simultaneous rotation in all directions, as in gliding upward, downward, forward, and backward.

The glenohumeral joint is an incongruous joint and requires all its components to function appropriately.

The glenoid fossa is shallow, and thus the humeral head "hangs" dependently and is not deep seated. The humerus does not "fall out" because of the muscular action of the supraspinatus muscle. The fossa, located at the outer angle of the scapula, faces forward, upward, and outward; thus, it supports the head of the humerus. The supraspinatus muscle and its tendon have limited elasticity, which thus prevents the head of the humerus from sliding down and away from the glenoid fossa. This muscle thus supports and moves the humeral head.

Abduction of the glenohumeral joint is accomplished by action of the deltoid muscle upon the humerus. As the line of force upon the humerus is upward elevation along the plane of the humerus, there must be simultaneous muscular action to initiate and assist in abducting the humerus. This muscular action is applied by contraction of the rotator cuff muscles upon the humerus, changing the angle of the humerus and thus permitting the deltoid to act at an angle and abduct as well as elevate the arm (Fig. 6–2).

The rotator muscles—the supraspinatus, infraspinatus, and teres minor muscles—act upon the head of the humerus in an eccentric manner via their conjoined tendon. This "off center" pull rotates the head of the humerus (Fig. 6–3), changes its angular relationship to the vertical axis, and allows the deltoid to become an abductor.

The humerus can be abducted by simultaneous rotator and deltoid muscle action to approximately 90 degrees of abduction. At 90 degrees of abduction, the humerus abuts upon the overhanging acromion and the coracohumeral ligament (Fig. 6–4, center drawing).

The major point of contact limiting abduction is the greater tuberosity of the humerus contacting the acromion. To some degree, there is a point of restricting contact of the greater tuberosity with the coracoacromial ligament.

With the arm internally rotated, the tuberosity contacts the acromion sooner and limits abduction by as much as 30 degrees (60 rather than 90 degrees of abduction). If the arm is externally rotated, the tuberosity moves behind the acromion and permits abduction to 120 degrees—an additional 30 degrees.

For full abduction in overhead elevation, the humerus must exter-

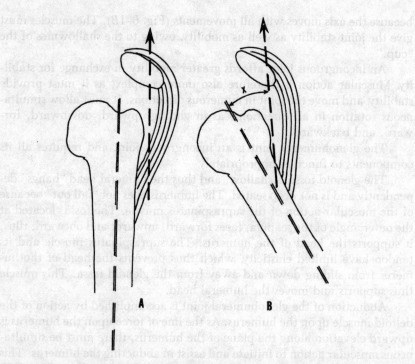

Figure 6–2. Abduction angle of deltoid. *A*, With the arm dependent, the deltoid line of pull is along the line of the humerus; thus the arm is elevated up against the acromion. *B*, With slight abduction of the humerus, the angle of the deltoid (X) changes its pull to abduction of the arm.

nally rotate and simultaneously abduct. This rotation is performed by the same muscles that abduct, the rotator cuff muscles (supraspinatus, infraspinatus, and teres minor muscles). They rotate about the axis of the humeral shaft as well as the axis of the humeral head (see Fig. 6–3).

With full abduction and simultaneous external rotation, the arm has still only reached 120 degrees of abduction and overhead elevation. For the arm to reach higher, the scapula must move. It must rotate about the chest wall to elevate the acromion further away from the greater tuberosity. The rotation of the scapula occurs with action correlating to the abduction and external rotation of the humerus, in what is termed **scapulohumeral rhythm.** This term indicates synchronous muscular contraction and free articular movements of the two incongruous joint surfaces.

The rotator cuff, the conjoined tendon of the four rotator muscles, is a major source of pain and disability. This tendon acts constantly in supporting the dependent arm in any abduction, forward flexion, or overhead elevation. In all movements, there is action between two opposing bony surfaces: the acromion and the greater tuberosity.

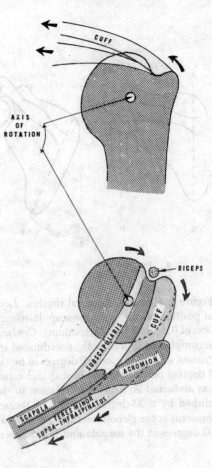

Figure 6-3. Rotator cuff attachment. *Top*, Abduction mechanism, AP view. *Bottom*, External rotation mechanism viewed from above.

This conjoined tendon, composed of parallel collagen fibers, has a limited blood supply. There is a zone proximal to its attachment to the greater tuberosity that is particularly devoid of blood supply. The vessels ascending from the circumflex artery of the humerus and those descending from the cuff muscles are compressed during the dependency of the arm and during any contraction of the cuff muscles. The tendon is thus ischemic for the greater portion of every day.

At times, there is an ample blood supply to this zone of the tendon—that is, when the arm is supported and neither dependent nor active.[1] This fluctuating circulation within the zone predisposes the patient to hyperemia and ischemia, with resultant pain from engorgement, and atrophy and degeneration from ischemia. The zone is aptly termed the "critical zone."

The capsule of the glenohumeral joint must be fully flexible. As the arm abducts, the inferior portion of the capsule must elongate to permit

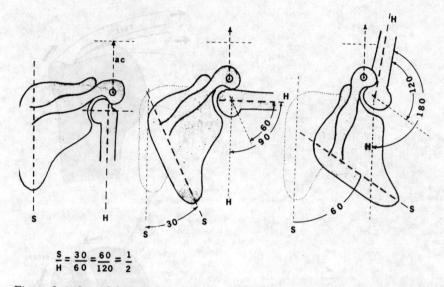

$$\frac{S}{H} = \frac{30}{60} = \frac{60}{120} = \frac{1}{2}$$

Figure 6–4. Scapulohumeral rhythm. *Left,* The scapula (S) and the humerus (H) at position of rest, with the scapula relaxed and the arm dependent, both at position of 0 degrees (ac—acromium). *Center,* The abduction movement of the arm is accomplished in a smooth, coordinated movement during which for each 15 degrees of arm abduction, 10 degrees of motion occurs at the glenohumeral joint, and 5 degrees occurs because of scapular rotation upon the thorax. The humerus (H) has abducted 90 degrees in relation to the erect body, but this has been accomplished by a 30-degree rotation of the scapula and a 60-degree rotation of the humerus at the glenohumeral joint, a ratio of 2:1. *Right,* Full elevation of the arm: 60 degrees at the scapula and 120 degrees at the glenohumeral joint.

full motion (Fig. 6–5). The superior aspect supports the dependent arm and must be intact and taut.

A bursa is located under the deltoid and the lateral aspect of the acromion with the inferior layer coating the inferior aspect of the cuff and the upper surface of the greater tuberosity and the humerus (Fig. 6–6). This bursa prevents excessive friction between the opposing surfaces of the acromion and the greater tuberosity and "protects" the tissues within the narrow suprahumeral space.

For the glenohumeral joint to be fully functional and painfree, the following conditions are necessary:

1. The musculature of the cuff and the deltoid must be well conditioned and be involved in a coordinated neuromuscular action. The movements of the scapula and humerus must always be coordinated.

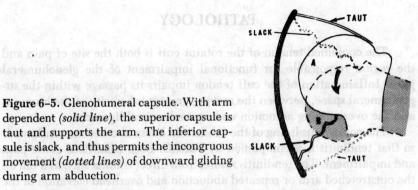

Figure 6-5. Glenohumeral capsule. With arm dependent *(solid line)*, the superior capsule is taut and supports the arm. The inferior capsule is slack, and thus permits the incongruous movement *(dotted lines)* of downward gliding during arm abduction.

2. The cuff tendon must be intact, with minimal thicking, inflammation, or degeneration.
3. The capsule must have adequate flexibility to allow full range of motion.
4. The subdeltoid bursa must be intact, well lubricated, and free of inflammation.
5. The articulating surfaces of the head of the humerus and the surface of the glenoid fossa must be smooth and well lubricated.

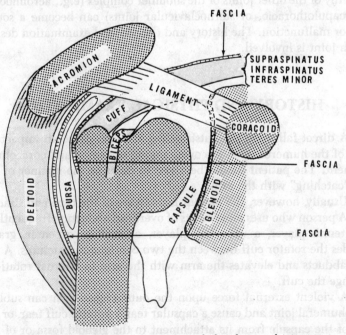

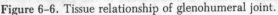

Figure 6-6. Tissue relationship of glenohumeral joint.

PATHOLOGY

The conjoined tendon of the rotator cuff is both the site of pain and the region responsible for functional impairment of the glenohumeral joint. Inflammation of the cuff tendon impairs its passage within the suprahumeral space, between the moving greater tuberosity of the humerus and the overhanging acromion and coracohumeral ligament.

The contiguous lining of the subdeltoid bursa also becomes inflamed, so that tendinitis is accompanied by bursitis. Both conditions cause pain and impairment. The tendinitis results from direct trauma such as a fall on the outstretched arm or repeated abduction and overhead elevation of the arm, especially when done without adequate external rotation.

Tearing of the collagen fibers of the rotator cuff tendon, either partial or complete, can cause pain and malfunction. Tearing of the tendon is possible, as the portion of the tendon entrapped between these two structures is the portion that has inadequate arterial circulation (the "critical zone").

Altered nerve supply to the muscles of the rotator cuff or the deltoid can cause malfunction and pain. These nerves are the suprascapular nerve to the rotator cuff and the axillary nerve to the deltoid. Injury of these nerves has numerous causes.

Any of the other joints of the shoulder complex (e.g., acromioclavicular, scapulothoracic, or sternoclavicular joints) can become a source of pain or malfunction. The history and the physical examination designates which joint is involved.

HISTORY AND PHYSICAL EXAMINATION

A direct fall upon the outstretched arm can cause an impact of the head of the humerus upon the overhanging acromion and coracohumeral ligament. The patient should be asked to describe the manner of falling and "catching" with the arm.

Usually, however, the patient is unaware of the type and time of injury. A person who uses movements of overhead abduction frequently, such as a tennis player, a volleyball player, or even a carpenter, gradually abrades the rotator cuff between the two contiguous structures. A person who abducts and elevates the arm with the arm in internal rotation can impinge the cuff.

A violent external force upon the outstretched arm can sublux the glenohumeral joint and cause a capsular tear, a rotator cuff tear, or even a tear of the capsule from its attachment to the glenoid fossa or of the labrum from the fossa.² Such a force can obviously occur during a contact sports activity, but it may also occur at a person's place of employment, during performance of a job-related task.

The physical examination findings depend on the extent of injury as well as the particular tissue involved. Usually, the patient presents with the following conditions:

1. There is tenderness over the precise site of injured tissue. Usually, this site is the greater tuberosity or the area immediately above the tuberosity and under the anterior margin of the acromion.
2. The arm is held supported, adducted, and internally rotated. This position relieves tension of the cuff tissues.
3. Abduction and external rotation, either active or passive, are restricted. As the glenohumeral joint does not move, any attempted motion of the arm is done with the scapula. Attempted abduction, forward flexion, and overhead extension are accomplished with a "shrugging mechanism" (Figs. 6–7 and 6–8). The attempted motion merely moves the scapula, and no glenohumeral motion occurs.

Either active or passive external rotation is painful and limited. If the rotator cuff is torn, there is *no* possible external rotation or abduction.

Injury to joints of the shoulder complex other than the glenohumeral joint will be separately discussed.

TREATMENT

Treatment of injury to the suprahumeral area and its tissues involves certain basic concepts and additional considerations for specific injuries. The basic concepts are as follows:

1. The glenohumeral joint should be rested for several days. To eliminate the effects of traction upon the arm, use of a sling for a few days may be valuable. This support also minimizes external rotation and abduction, both of which are painful and irritating to the injured tissues.
2. Ice packs should be applied for 20 minutes several times a day.
3. Oral anti-inflammatory medications that are not contraindicated for the particular patient should be administered.
4. Pendular exercises should be started within a few days to prevent adhesions from forming in the glenohumeral capsule or the subdel-

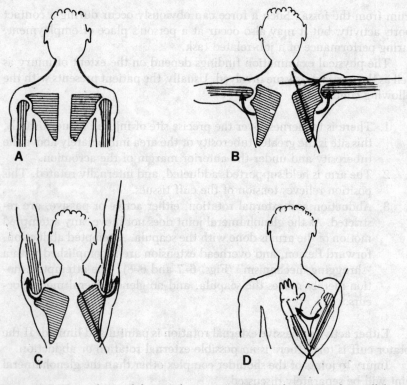

Figure 6–7. Normal scapulohumeral motions. *A,* Normal stance with parallel symmetric shoulder girdles. *B,* Symmetric abduction with equal proportional glenohumeral abduction and scapular rotation. *C,* Symmetric full overhead arm elevation. *D,* Posterior arm flexion and internal rotation.

toid bursa. These exercises must be performed precisely and frequently to prevent adhesions (Fig. 6–9).

5. A solution containing an analgesic and steroid should be injected into the suprahumeral area, the area of the rotator cuff. This injection decreases the pain and the inflammation. The value of adding steroid to the injection has been questioned, but its clinical benefit has been demonstrated. The site of injection is the suprahumeral area directly under the anterior border of the acromion (number 4 in Fig. 6–10). When the space between the acromion and the greater tuberosity has been palpated, the needle enters and slowly proceeds superiorly (Fig. 6–11). After aspiration is performed to determine any vascular entry, a solution of several milliliters of analgesic agent and 1 mm of soluble steroid is injected.

6. Active exercises should be gradually instituted. Forward flexion is the easiest and is followed by posterior flexion. External rotation is

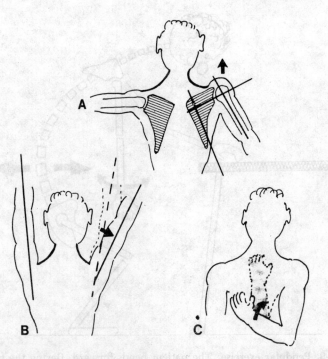

Figure 6–8. Subtle signs of shoulder limitation. *A*, Shrugging with excessive scapular rotation and limited glenohumeral abduction. *B*, Limited right arm overhead elevation. Arm "away" from head and ear. *C*, Limited posterior flexion and internal rotation. Hand fails to reach normal interscapular distance of reach.

more difficult and more painful, yet very important (Fig. 6–12). Gradually, active posterior flexion is encouraged. Figure 6–13 provides an example of this type of exercise. There are many other exercises using posterior flexion, one of which involves bringing the arm backward, using the good arm to increase the range.

Abduction should be avoided until the arm is reasonably pain-free and all other motions are possible. Abduction tends to cause the greater tuberosity to encroach up and under the acromion, thus maintaining inflammation and causing further damage to the tendon.

THE SUBTLE SIGNS
OF RESIDUAL PERICAPSULITIS

Often, patients who have diligently adhered to all aspects of prescribed treatment and who assume that normal range of motion has been

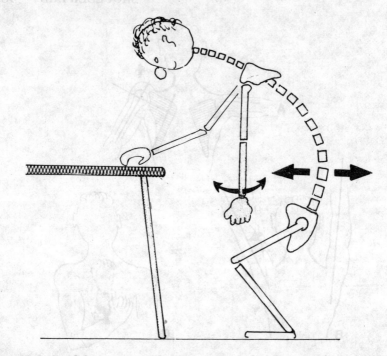

Figure 6–9. Pendular exercise. The patient bends forward, flexing the trunk to right angles. The involved arm is dangled without muscular activity of the glenohumeral joint. The body actively sways, thus passively swinging the dependent arm in forward flexion-extension, lateral swing, and rotation. The body can be supported by placing the other arm upon a table or chair.

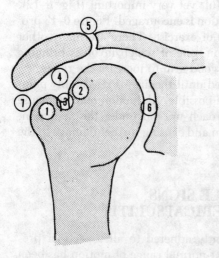

Figure 6-10. Sites of tissue pain: *1*, Greater tuberosity: attachment of supraspinatus tendon. *2*, Lesser tuberosity. *3*, Bicipital groove: tendon of long head of biceps. *4*, Subacromial bursa. *5*, Acromioclavicular joint. *6*, Glenohumeral joint and capsule. *7*, Subdeltoid bursa. (Modified from Steindler, A: The Interpretation of Pain in Orthopedic Practice. Charles C Thomas, Springfield, IL, 1959.)

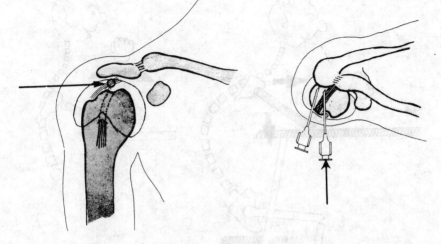

Figure 6–11. Site of injection in pericapsulitis. *Left*, Region of supraspinatus insertion in the suprahumeral space. The region is palpable immediately below the overhanging acromion and by palpation of the greater tubersotiy just lateral to the bicipital groove of the humerus. *Right*, Insertion of needle viewed from above. Two directions of entrance are shown, with the arrow depicting that shown in the anterior view.

restored despite some lingering pain demonstrate subtle signs of residual pericapsular limitation on careful examination (Fig. 6–14).

Overhead elevation of the injured arm lacks several degrees of range as compared with that of the normal arm. Posterior flexion and internal rotation (reaching behind one's back) are limited. Movement is also limited in trying to bring the elbows back with hands behind the head. External rotation with the arm dependent (elbows at patient's side) is a subtle sign of residual pericapsulitis. "Shrugging" that is not apparent to the patient and that occurs on request of abduction and overhead elevation is another sign.

When these subtle signs of residual pericapsulitis are discovered, physical therapy for passive and active assisted exercises must be added to the prescribed exercise regimen. These additional exercises may involve rhythmic stabilization, contraction and relaxation, or passive stretching while a device for transcutaneous electrical nerve stimulation (TENS) is applied. Application of ice and active stretching are also of value.

A series of suprahumeral injections are of value to patients with residual pericapsulitis. Mobilization or gentle manipulation may be needed. Suprascapular nerve blocks are valuable (Fig. 6–15).

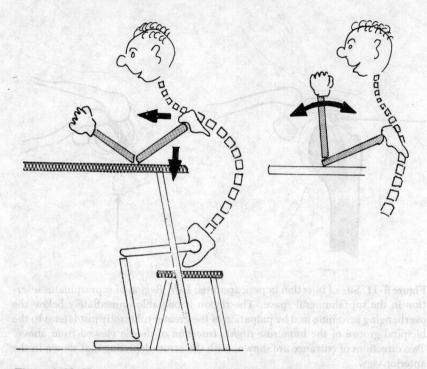

Figure 6-12. Home exercise to increase shoulder range of motion. Seated with arm supported upon table, the patient moves forward and downward to increase range of arm toward elevation. The forearm, bent at right angles, internally and externally rotates, thus further increasing range. A weight can be held in the hand during this exercise.

FROZEN SHOULDER

The term **frozen shoulder** refers to the glenohumeral joint that is incapable of passive or active movement. The joint cannot move or be moved in abduction, forward or backward flexion, or external rotation.

The frozen shoulder is a sequela of acute or chronic pericapsulitis following tendinitis. It may also be a sequela of myocardial infarct, dislocation, prolonged immobilization, or even cervical radiculitis.

The condition is actually a combination of adhesive capsulitis, adhesive subdeltoid bursitis, and adhesive biceps tendinitis. The exact tissue mechanism is equivocal, but it involves adhesion of the contiguous synovial layers of the capsules (Fig. 6-16). Inflammation of these synovial layers causes an outpouring secretion of exudate that contains protein. Microscopic fibers attach from adjacent synovial layers, which then multiply,

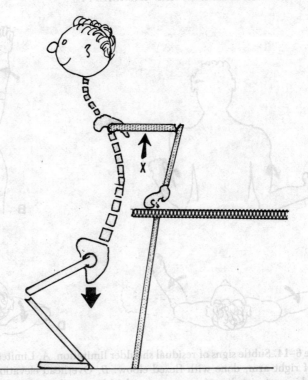

Figure 6-13. Exercise to stretch anterior capsule and increase posterior flexion. The patient places both hands on a table behind the back and performs gentle deep knee bends. This exercise elevates arms *(small arrow)* and increases posterior flexor range of motion.

thicken, and shorten. A sympathetic component is considered to exist with vasospasm and formation of mast cells into fibroblasts (Fig. 6–17).

Clinically, there is progressive limitation of range of motion *in all directions.* Pain may or may not exist, but all motion attempted beyond the possible range causes capsular pain. The condition usually progresses until *all* motion is restricted, at which time there is no pain.

The primary treatment is prevention. Recognition of this condition in its early phase is mandatory. For patients with "low pain threshold," excessive vasomotor instability, excessive emotional lability, and a passive attitude to treatment, active aggressive treatment must be instituted to regain full range of motion. In the early stages, the condition is reversible, but once there is fibrous adhesion, some permanent restriction remains.

All the modalities described must be instituted. In addition, antidepressant drugs, for reasons not completely understood at present, have proved to be valuable. The use of a stellate ganglia block (Fig. 6–18) is also worth considering, as is the use of oral steroids.

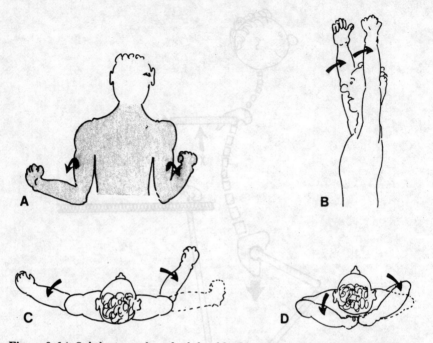

Figure 6–14. Subtle signs of residual shoulder limitation. *A*, Limited external rotation of right arm, done with flexed elbow. *B*, Overhead elevation of right arm limited in posterior direction as compared with normal arm, viewed from side. *C*, External rotation as viewed from above. *D*, With hands behind head, right arm fails to fully extend posteriorly.

ROTATOR CUFF TEAR

As previously stated, the conjoined tendon of the rotator cuff is a relatively ischemic tissue within the "critical zone." With progressive deterioration from excessive use and abrasion or following an external injury, the tendon may tear. In a tear, there is disruption of the continuity of the parallel collagen fibers. After the age of 40, all rotator cuffs are claimed to show some degree of deterioration. A tear may be complete or incomplete: all fibers may lose their continuity, or merely a portion of the fibers may be disrupted.

Complete Rotator Cuff Tear

When the rotator cuff tendon is completely torn, immediate pain similar to that of any rotator cuff injury may be experienced. Immediately

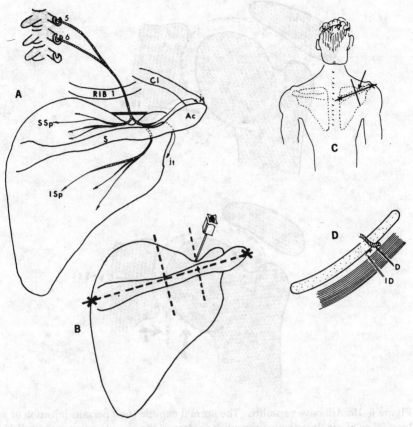

Figure 6-15. Site of suprascapular nerve block. *A,* Normal anatomy of suprascapular nerve. Cl = clavicle. Ac = acromium. S = scapular spine. Jt = joint (glenohumeral). SSp = supraspinatus. Isp = infraspinatus. *B,* Surface anatomy for site of injection. Spine of scapula is measured, and midpoint is estimated. *C,* Nerve is halfway between midpoint and lateral tip of spine. *D,* Direct (D) and indirect (ID) needle entrances to the nerve. In indirect approach of anesthetic agent to nerve, needle is slightly distant to nerve emergence.

with complete tear, the person becomes unable to abduct the arm and elevate it overhead. The "shrugging mechanism" described in the section on residual pericapsulitis is also exhibited in complete rotator cuff tear as a result of pain from entrapment of the tendon between the acromion and greater tuberosity.

Because of the complete tear, the arm cannot be actively abducted, but once it is passively abducted, it can be held in that position by the patient, although only briefly. The arm then slowly descends to the side position.

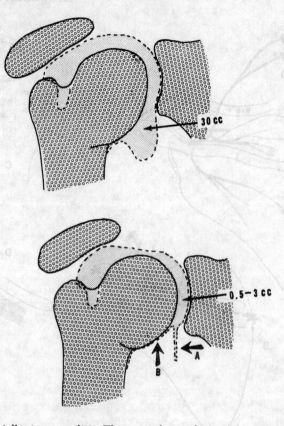

Figure 6-16. Adhesive capsulitis. The normal capsule *(top)* permits injection of at least 30 cc of air. In adhesive capsulitis *(bottom)*, the capsule adheres to itself (A) and to the humeral head (B), which decreases capacity to 0.5 to 3 cc and markedly limits range of motion.

The reason that the arm cannot be actively abducted is that the humeral head cannot be externally abducted, owing to the rotator cuff's being "disconnected." Thus, the humerus cannot be abducted to the angle that allows the deltoid to abduct the arm further. The reason that the humerus can be held once it is passively abducted is that the examiner performs the function of the rotator cuff and places the humerus at an angle that normally permits the deltoid to hold the arm in active abduction.

The reason that the arm cannot be held indefinitely in an abducted position or against any downward pressure is that the rotator cuff cannot "seat" (hold) the head of the humerus into the glenoid fossa; thus, the weight of the arm overwhelms the supporting musculature, and the arm "drops."

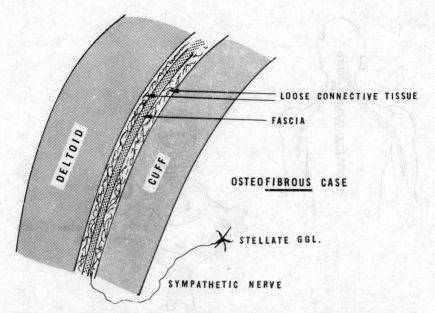

Figure 6-17. Schematic of the osteofibrous case of the glenohumeral joint. The tissue relationship of the fascia that is between the rotator cuff and the undersurface of the deltoid is *loose* connective tissue to permit gliding. Adhesion occurs when there is thickening of the connective tissue between the two surfaces.

In a complete tear, the arm cannot voluntarily be externally rotated. This can be tested by having the patient place the affected arm in the dependent position with the elbow flexed 90 degrees. The person is asked to turn the hand "out," which externally rotates the humerus. Failure to externally rotate the arm (against some resistance from the examiner) indicates a torn rotator cuff.

A complete tear can usually be clinically diagnosed with certainty. If necessary, arthrography may be used. In an arthrogram, the dye injected into the glenohumeral capsule is seen to "leak" through the tear into the subdeltoid bursa (Fig. 6–19). Although the completeness of the tear cannot be ascertained solely on the basis of arthrography, a "positive" arthrogram can lend support to the clinical diagnosis.

In the acute phase of a tear, the patient's pain may prevent an accurate diagnosis. The pain can be relieved by suprahumeral injection of an anesthetic agent. Once the pain has diminished, the strength or absence of external rotation can be ascertained.

Treatment. A complete tear does not heal spontaneously; thus, any physical therapy is palliative for pain but does not eliminate dysfunction. Surgical repair is required and must be performed early, before the proximal portion of the rotator cuff retracts (within the first week or sooner).

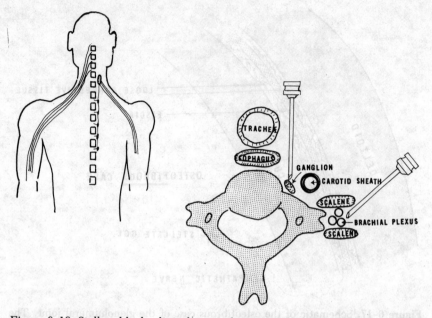

Figure 6–18. Stellate block: clinical sympathectomy. The block is performed from the anterior neck approach, slightly lateral to the trachea with the needle penetrating to the vertebral body. Usually 8 ml of 1 percent procaine is effective within 10 minutes.

Partial Rotator Cuff Tear

A partial cuff tear occurs after trauma more frequently than is usually suspected. After the age of 35, the rotator cuff undergoes degenerative changes. These changes are more apparent in physically active people who frequently place their arms overhead with or without rotation or who hold their arms abducted or in forward flexion for sustained periods.

In a partial cuff tear, there are sufficient intact fibers remaining to abduct and externally rotate the arm. The fibers that tear, however, tend to retract and bunch up, causing a "tumor" at that site of the cuff (Fig. 6–20). When the bunched fibers pass under the acromion and the coracohumeral ligament during abduction and overhead elevation, the arm may exhibit the "painful arc." There may be pain or limited motion similar to that noted in rotator tendinitis.

The "shrugging mechanism" is present. The major diagnostic difference between tendinitis and partial tear is the prolongation of symptoms despite adequate therapy.

Treatment. Therapy is the same as that for tendinitis. The presence of the tear may be confirmed by arthrography. If the tear is large, surgical repair may be necessary, as not even a partial tear heals spontaneously. The

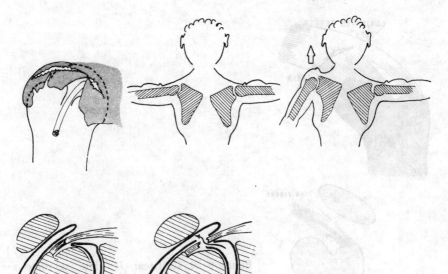

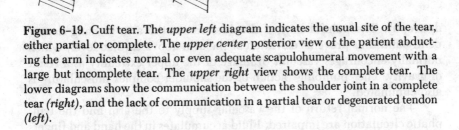

Figure 6–19. Cuff tear. The *upper left* diagram indicates the usual site of the tear, either partial or complete. The *upper center* posterior view of the patient abducting the arm indicates normal or even adequate scapulohumeral movement with a large but incomplete tear. The *upper right* view shows the complete tear. The lower diagrams show the communication between the shoulder joint in a complete tear *(right)*, and the lack of communication in a partial tear or degenerated tendon *(left)*.

scar reaction of the tear may cause persistent "subtle" signs of shoulder tendinitis.

Many clinicians advocate early surgical intervention to remove the damaged tissue and repair the tear. When a "painful arc" persists, resection of either the anterolateral tip of the acromion, the coracoacromial ligament, or both may be considered.

Regardless of the "definitive" treatment ultimately undertaken, it is mandatory that the full range of glenohumeral joint motion be regained and that the rotator muscles be strengthened.

SHOULDER-HAND-FINGER
SYNDROME (SHFS)

Any painful limited glenohumeral shoulder can develop into shoulder-hand-finger syndrome (SHFS), a variant of reflex sympathetic

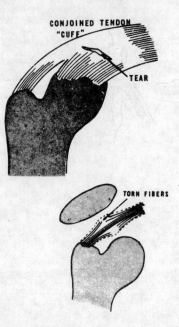

Figure 6-20. Cuff tear. *Top*, Site and direction of partial cuff tear. *Bottom*, Retraction of torn cuff fibers forms a thickening of the cuff, thus resembling the thickening of tendinitis.

dystrophy. The painful limited shoulder causes the arm to hang dependently and thus prevents elevation of the arm above cardiac level, which normally encourages vascular drainage of the upper extremity as a result of gravity. The "pumping" action of the shoulder complex musculature is also eliminated.

The normal return of the vascular supply to the arm and the lymphatic circulation are impaired. Fluid accumulates in the hand and fingers because of elevated venous pressure and venous distension. A surplus of lymphatic fluid and edematous fluid also accumulate in the hand.

As the major venous supply and lymphatic supply of the finger are on the dorsum, swelling occurs there first. The normal wrinkles of the fingers can be noted. The fingers become puffy and pale and exhibit pitting edema. They can no longer be fully flexed into a clenched fist; thus, the pumping action of the hand becomes diminished, and more edema results. Gradually, all three stages of SHFS occur, with or without pain.

The best and only effective treatment for SHFS is prevention of progression of the early stage into organized edema. Such prevention requires that the patient regains full range of shoulder motion and performs full-range exercises of the hand. The shoulder exercises have been discussed previously in this chapter.

The hand, especially the fingers, can be "milked" of the accumulated fluid by compressive wrapping with heavy string, twine, or thin rope, applied distally to proximally. The arm must be held in an elevated position—above the heart level—during major portions of the day and night. The hand must be clenched into a full fist position repeatedly during the day.

Stellate blocks are usually advocated for causalgia and pain, but even when the patient lacks the characteristic pain of SHFS, the use of blocks and oral steroids has been advocated.

ACROMIOCLAVICULAR JOINT PAIN

Another joint in the shoulder complex that can frequently cause pain and impairment is the **acromioclavicular (AC) joint**. The commonest cause of AC joint impairment is direct trauma: a force from above or from the front.

Normally in scapulohumeral abduction and overhead elevation, the scapular phase rotates about the clavicle at the AC joint. Movement of the clavicle occurs at the AC joint and about the sternoclavicular joint. The AC joint is not a true joint but a synarthrosis and must rotate about its long axis when it moves to allow elevation of the scapula. The AC joint has a pseudomeniscus that permits movement.

Pathology resulting from a direct blow frequently causes AC separation of varying degree. The resultant articular damage results in localized tenderness, pain at that joint on shoulder movement, and often, visible asymmetry. Auscultation may reveal crepitation during movement, which can be heard using a direct ear position or a stethoscope.

The history of a direct injury should be elicited, and the findings of localized pain noted. To differentiate AC joint pain from glenohumeral joint pain, the physician should attempt to reproduce AC pain by having the patient elevate the scapula ("shrug") without abducting the arm.

Local injection of several milliliters of an anesthetic agent into the joint confirms the diagnosis. X-ray studies of the arm in a dependent position and holding a heavy weight should reveal a separation.

Treatment of an AC separation was originally surgical repair, but more recently conservative management has been advocated, and the results have been gratifying. The arm is elevated and held backward by an appropriate sling or figure-of-8 bandage. Injection of local analgesic relieves the pain during healing.

In the case of complete tear of the coracoclavicular ligament, surgical repair is indicated.

SCAPULOCOSTAL SYNDROME

In scapulocostal syndrome, the patient claims to experience pain in the interscapular area, usually with muscular tenderness and even "nodularity" of the rhomboids, levator scapulae, and trapezius muscles.

In normal shoulder abduction and overhead elevation, the scapula rotates about the clavicle and upon the posterior rib cage. The scapula rotates forward by contraction of the serratus muscle and is elevated by contraction of the trapezius muscle. The rhomboid muscles depress the scapula and cause it to rotate downward.

Any arm movement away from the body requires contraction of the scapular muscles to support the extended arm. With sustained elevation of the arm, the scapular muscles maintain isometric contraction. Overuse of these muscles can cause fatigue and subsequent pain.

Poor posture with excessive dorsal kyphosis places a direct strain on the scapular muscles. Anxiety also causes the scapular muscles to undergo fatiguing isometric contraction with subsequent muscular pain. Occupational stresses, such as prolonged leaning over a computer, aggravates or causes scapulocostal muscular pain.

Recent studies using electromyographic (EMG) needles in the trapezius, rhomboid, and levator scapulae muscles during positions of forward arm extension have revealed fatigue of those muscles. Fatigue occurs within 8 to 15 minutes of sustained forward arm postures. Ischemia of these muscles also occurs from prolonged isometric muscular contraction.

The tender "nodules" that develop within these muscles constitute one of the findings of fibrositis, fibromyositis, and so-called "trigger points."

Treatment of scapulocostal syndrome is (1) improvement of posture, (2) decrease of forward head posture, (3) correction of occupational postural stresses, (4) elimination of the factors of tension, anxiety, fatigue, depression, or anger that influence the posture, (5) local injections into the trigger areas with an anesthetic agent, and (6) postural exercises combined with application of ice or vasocoolant spray to the muscles undergoing tension and passive stretching of the involved muscles.

SHOULDER (GLENOHUMERAL) DISLOCATION

The glenohumeral joint can be dislocated by an excessive external force that subluxes the joint and stretches or tears the capsule. The glenohumeral joint is well protected anteriorly by the pectoral muscles, superiorly by the acromion and the deltoid muscle, and posteriorly by the deltoid and latissimus muscles. The inferior portion of the joint represents a "weak" spot because it is protected only by the capsule. Inferior dislocation is thus the most common site of shoulder dislocation.

A fall on the outstretched arm causes a leverage stress upon the joint with inferior dislocation (Fig. 6–21). Any fall that forces the arm backward from the position of forward flexion and elevation can cause dislocation.

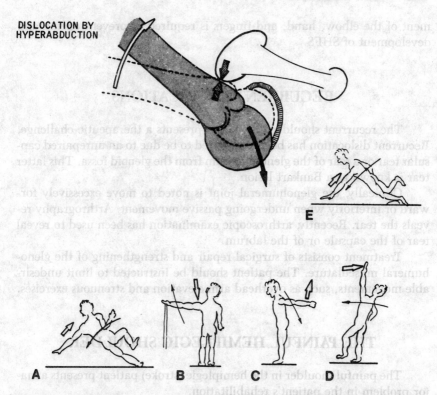

DISLOCATION BY
HYPERABDUCTION

Figure 6-21. Mechanism of dislocation: "hyperextension theory." Abduction with the humerus in internal rotation or forward flexion with the arm in external rotation becomes limited by the acromial arch. Forceful elevation when this point of impingement has been reached uses the arch as a fulcrum and dislocates the proximal head by causing it to descend and move forward. The *arrows* in A through E indicate the external forces that can cause dislocation of the glenohumeral joint.

Diagnosis is usually obtained clinically from eliciting the history and noting the appearance of the shoulder and the severe limitation, both active and passive, of glenohumeral movement. Verification by x-ray studies completes the diagnosis.

Treatment requires relocation of the head of the humerus within the glenoid fossa. This relocation can be accomplished by using the Kocher maneuver or by using traction upon the arm with a counter force into the axilla (pulling the arm gently and constantly in traction with the foot against the armpit).

After reduction of the dislocation, the arm should be splinted in the adducted internally rotated position for a period of approximately 6 weeks. When the arm is in this splinted position, constant active movement of the elbow, hand, and fingers is required to prevent edema and development of SHFS.

RECURRENT DISLOCATIONS

The recurrent shoulder dislocation presents a therapeutic challenge. Recurrent dislocation has been considered to be due to an unrepaired capsular tear or a tear of the glenoid labrum from the glenoid fossa. This latter tear is known as a Bankart lesion.

Clinically, the glenohumeral joint is noted to move excessively forward or inferiorly when undergoing passive movement. Arthrography reveals the tear. Recently, arthroscopic examination has been used to reveal tear of the capsule or of the labrum.

Treatment consists of surgical repair and strengthening of the glenohumeral musculature. The patient should be instructed to limit undesirable movements, such as overhead arm elevation and strenuous exercises.

THE PAINFUL HEMIPLEGIC SHOULDER

The painful shoulder in the hemiplegic (stroke) patient presents a major problem in the patient's rehabilitation.

In the normal patient, the glenohumeral joint remains well supported by (1) good tone and limited elongation of the supraspinatus muscle and (2) maintenance of the angulation (in forward, upward, and outward directions) of the glenoid fossa. The second means of support depends on the scapular musculature and the erect posture of the spine.

During the acute flaccid stage of hemiplegia, the patient's spine laterally deviates, and the scapula rotates downward, as a result of temporary flaccidity of the scapular musculature (Fig. 6–22). The head of the humerus being abducted rotates downward and outward and subluxes (Fig. 6–23).

The duration of the flaccid stage of hemiplegia varies; it may involve minutes, weeks, or months. During this stage, the scapula rotates outward, causing the humerus to sublux and stretch the capsule. At this stage, the capsule is the sole supporter of the glenohumeral joint, as the rotator cuff is paretic. The supraspinatus muscle does not prevent downward subluxation.

Any movement of the arm can cause further subluxation, stretching, or tearing, of the capsule. Such movements may be the patient's active changes of position in bed or in a wheelchair, or passive movements of the patient by a nurse or relative.

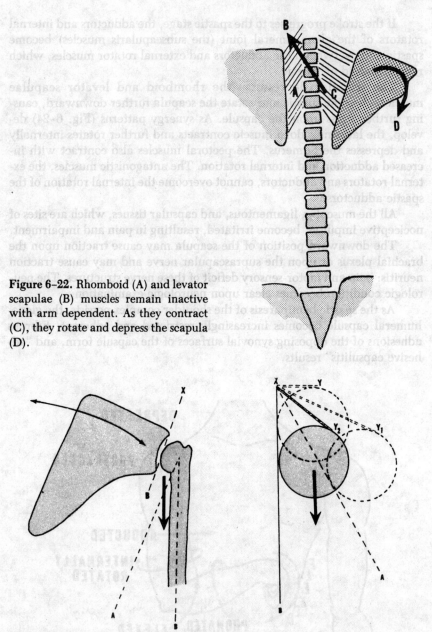

Figure 6-22. Rhomboid (A) and levator scapulae (B) muscles remain inactive with arm dependent. As they contract (C), they rotate and depress the scapula (D).

Figure 6-23. With downward rotation of the scapula, the angle of the glenoid fossa (X-B) changes, causing the arm to abduct (B). The humerus subluxes downward *(arrow)*. The rotator cuff muscles—the supraspinatus, infraspinatus, and teres minor muscles (Y), which normally prevent downward rotation, no longer support the humerus.

If the stroke progresses to the spastic stage, the adductors and internal rotators of the glenohumeral joint (the subscapularis muscles) become spastic and overwhelm the abductors and external rotator muscles, which remain flaccid or paretic.

The scapular depressors—the rhomboid and levator scapulae muscles—become spastic and rotate the scapula further downward, causing further stress upon the capsule. As synergy patterns (Fig. 6–24) develop, the latissimus dorsi muscle contracts and further rotates internally and depresses the humerus. The pectoral muscles also contract with increased adduction and internal rotation. The antagonistic muscles, the external rotators and abductors, cannot overcome the internal rotation of the spastic adductors.

All the muscular, ligamentous, and capsular tissues, which are sites of nociceptive impulses, become irritated, resulting in pain and impairment.

The downward position of the scapula may cause traction upon the brachial plexus or upon the suprascapular nerve and may cause traction neuritis: pain and motor-sensory deficit of these nerve structures. The neurologic condition becomes clear upon neurologic examination.

As the spastic hemiparesis of the shoulder complex persists, the glenohumeral capsule becomes increasingly restricted and adherent. Fibrous adhesions of the opposing synovial surfaces of the capsule form, and "adhesive capsulitis" results.

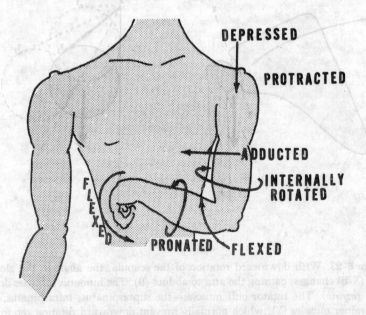

Figure 6–24. The spastic hemiplegic shoulder.

In addition to capsular limitation, adhesive fibrous reaction occurs in the subscapular bursa. Joint limitation now results from adhesive glenohumeral capsulitis and adhesive subdeltoid bursitis. A "frozen shoulder" evolves.

A painful shoulder in the hemiplegic patient may occur from any of the preceding mechanisms. As a result of the painful shoulder and progressively limited movement, reflex sympathetic dystrophy may evolve. The vasomotor changes of the upper extremity develop as in any SHFS.

The mechanisms implicated provide a basis for preventative measures in rehabilitating the hemiplegic patient. They also indicate the therapeutic requirements.

Positioning of the acute flaccid hemiplegic patient mandates postural correction of the trunk to prevent lateral "leaning," the functional scoliosis. Maintenance of scapular elevation and retraction is also required. "Splinting" of the shoulder refers to application of these principles. The head of the humerus must be elevated into the acromion, the coracohumeral ligament, and the glenoid fossa. Laying on the hemiplegic side should not be avoided. Leaning upon the abducted extended upper extremity in the sitting position should be encouraged.

Slight external rotation of the arm and elevation of the scapula are indicated. In cases of total flaccid paresis, use of electrical stimulation is valid until some reflex or voluntary motor function returns. This electrical stimulation applies to the supraspinatus, deltoid, and trapezius muscles.

As spasticity evolves into the synergy pattern, the uninhibited subscapular muscle, which is an internal rotator, must be judiciously stretched and kept elongated until the external rotators and abductors regain some control. The latissimus dorsi muscle must also be viewed as an internal rotator that becomes contracted, and it too must be kept under control passively until there is a return of active balance.

Each stage of hemiplegia—flaccidity, spasticity, and synergy—involves separate mechanisms of pain and dysfunction, and these mechanisms must be always recognized and minimized. The "frozen shoulder," rotator cuff tear, reflex sympathetic dystrophy, and other complications add a formidable adverse component to the rehabilitation of the hemiplegic patient.

REFERENCES

1. Mosele, FH, and Goldie, I: The arterial pattern of the rotator cuff of the shoulder. J Bone Joint Surg 45-B:780–789, 1963.
2. Bankart, ASB: The pathology and treatment of recurrent dislocation of the shoulder joint. Br J Surg, Dec 15, 1923, pp 23–29.

BIBLIOGRAPHY

Bankart, ASB: The pathology and treatment of recurrent dislocation of the shoulder joint. Br J Surg, Dec 15, 1923, pp 23–29.

Bjelle, A, Hagsberg, M, and Michaelsson, G: Clinical and ergonomic factors in prolonged shoulder pain among industrial workers. Scand J Work Environ Health 5:205–210, 1979.

Cailliet, R: Shoulder Pain, ed 2. FA Davis, Philadelphia, 1981.

Jonsson, B, and Hagberg, M: Vocational electromyography in shoulder muscles in an electronic plant. In Biomechanics, vol. 7. University Park Press, Baltimore, 1987.

Mennell, JM: Joint Pain. Little, Brown & Co, Boston, 1964.

Mosele, FH, and Goldie, I: The arterial pattern of the rotator cuff of the shoulder. J Bone Joint Surg 45-B:780–789, 1963.

Neer, CS, and Marberry, TA: On the disadvantages of radical acromionectomy. J Bone Joint Surg 63-A:416–419, 1981.

Neviaser, JS: Adhesive capsulitis of the shoulder. J Bone Joint Surg 27:211–222, 1945.

Neviaser, RJ: Treating patients with rotator cuff tears. Journal of Musculoskeletal Medicine, April 1985, pp 17–23.

Rizk, TE, Christopher, RP, Pinals, RS, and Frix, R: Adhesive capsulitis (frozen shoulder): A new approach. Arch Phys Med Rehabil 64:29–33, 1983.

Rosenberg, PS, and Clarke, RP: Chronic rotator cuff tears: A review paper. Orthopedic Review 15:33–42, 1986.

Rubin, D: An exercise program for shoulder disability. Calif Med 106:39–43, 1967.

Subbarao, J, Stillwell, GK: Reflex sympathetic dystrophy syndrome of the upper extremity: Analysis of total outcome of management of 125 cases. Arch Phys Med Rehabil 62:549–554, 1981.

Weiss, JJ: Intraarticular steroids in the treatment of rotator cuff tear: Reappraisal by arthrography. Arch Phys Med Rehabil 62:555–557, 1981.

Welfling, J: Painful Shoulder—Frozen Shoulder. Falio Rheumatologia, Geigy Pharmaceuticals, New York, 1969.

CHAPTER 7

Elbow Pain

FUNCTIONAL ANATOMY

The elbow joint consists of three articulations: the **humeroulnar,** the **capitular radial** (radiohumeral), and the **radioulnar** joints (Fig. 7–1). The humeroulnar joint permits flexion and extension of the arm, and the other two joints permit pronation and supination of the forearm.

The humeroulnar joint determines the carrying angle of the elbow. When the arm is flexed, the forearm is in line with the upper arm. As the arm extends, the trochlea causes valgus of the forearm. Pronation and supination of the forearm occur about the capitular radial and the radioulnar joints.

There are two major collateral ligaments that stabilize the elbow: an anterior band and a posterior band. The anterior band arises from the medial epicondyle and attaches to the medial side of the coronoid process of the ulna. It is the major stabilizing ligament of the elbow. The posterior band is thinner and restricts the elbow when the elbow is flexed to an angle greater than 90 degrees.

The following muscles act upon the elbow:

1. The **brachialis,** which originates from the lower aspect of the lower half of the humerus and attaches to the anterior aspect of the coronoid process. It is the major flexor of the elbow.
2. The long and short heads of the **biceps,** which unite about the middle of the humerus and insert upon the medial aspect of the radius. This muscle flexes, but predominantly supinates, the forearm.

209

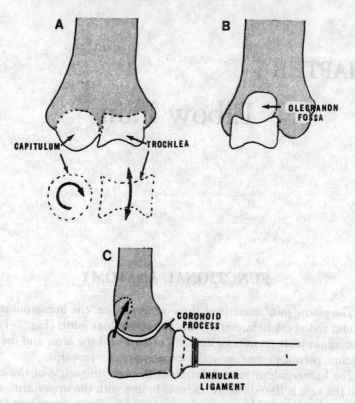

Figure 7-1. Elbow joint bony anatomy. *A*, Anterior view depicting the round sphere of the capitulum, upon which the radius rotates, and the spool-shaped trochlea, about which the ulna flexes and extends. *B*, Posterior view of the humerus showing the olecranon fossa, into which the posterior (olecranon) portion of the radius enters upon elbow extension. *C*, Lateral view of the elbow joint.

   3. The **triceps,** which originates from the lower posterior aspect of the humerus and inserts upon the ulna. This muscle extends the arm.

The flexor carpi radialis muscle, palmaris longus muscle, and parts of the flexor digitorum sublimis muscle originate from the medial epicondyle. One head of the pronator teres muscle originates from the epicondyle. This flexor forearm muscle group exerts a powerful pull upon the epicondylar periosteum (Fig. 7-2).

The extensor forearm muscle group arises from the lateral epicondyle of the humerus and extends to the wrist and fingers (Fig. 7-3).

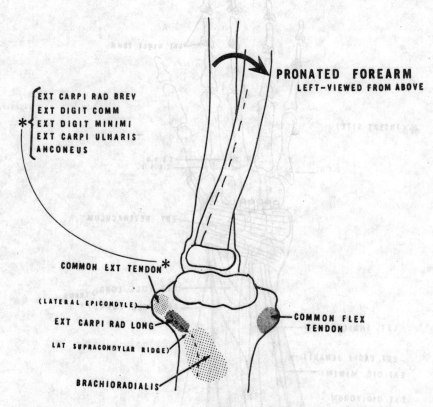

PRONATED FOREARM
LEFT-VIEWED FROM ABOVE

[EXT CARPI RAD BREV
EXT DIGIT COMM
*{EXT DIGIT MINIMI
EXT CARPI ULNARIS
ANCONEUS

COMMON EXT TENDON *

(LATERAL EPICONDYLE)

EXT CARPI RAD LONG

LAT SUPRACONDYLAR RIDGE)

BRACHIORADIALIS

COMMON FLEX
TENDON

Figure 7-2. Bony aspect of elbow in pronated position. Site of muscular attachment of forearm muscles.

In the articular motion of the elbow when the elbow is fully extended, the olecranon tip enters the olecranon fossa. The fossa must be completely open to receive the tip and allow full extension of the arm.

Examination of the elbow must include both active and passive ranges of motion. Active flexion-extension (humeroulnar joint) and active pronation-supination (radioulnar joint) must be tested.

Passive examination reveals the integrity of both joints, including the flexibility of the periarticular soft tissues. Active range of motion with or without resistance is tested to evaluate the strength and endurance of the related muscular tissues.

Active flexion, extension, pronation, and supination are tested to evaluate the adequacy and integrity of the nerve supply to the involved muscles. The peripheral nerves of the brachial plexus and the cervical cord root levels are thus evaluated.

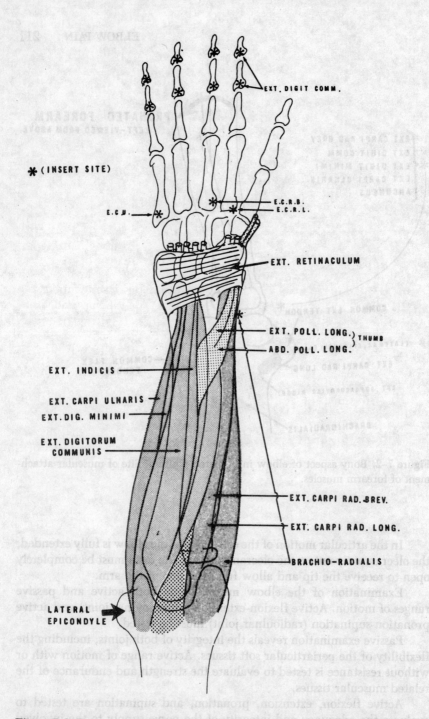

EXT. DIGIT COMM.

* (INSERT SITE)

E.C.U. →

E.C.R.B.
E.C.R.L.

EXT. RETINACULUM

EXT. POLL. LONG. } THUMB
ABD. POLL. LONG.

EXT. INDICIS

EXT. CARPI ULNARIS

EXT. DIG. MINIMI

EXT. DIGITORUM
COMMUNIS

EXT. CARPI RAD. BREV.

EXT. CARPI RAD. LONG.

BRACHIO-RADIALIS

LATERAL
EPICONDYLE

Figure 7-3. Extensor aspect of forearm. The extensor group is shown with particular emphasis on the origin of the common extensor group from the lateral epicondyle. Muscles *marked with asterisks* are innervated by the radial nerve.

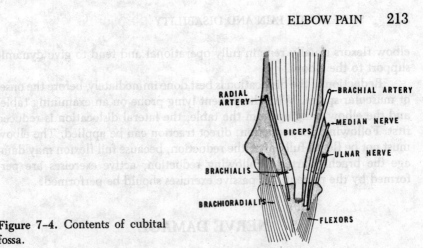

RADIAL ARTERY

BRACHIAL ARTERY

MEDIAN NERVE

BICEPS

ULNAR NERVE

BRACHIALIS

BRACHIORADIALIS

FLEXORS

Figure 7-4. Contents of cubital fossa.

The anterior aspect of the joint, the antecubital fossa, contains the biceps tendon, the radial and brachial arteries, the median and ulnar nerves, and the origin of many of the forearm muscles (Fig. 7-4). All these tissues are available to examination.

TRAUMA

Trauma to the elbow may impair any of the three joints with limited flexion and extension, or pronation and supination. Pronation and supination may exist in the absence of flexion and extension, and vice versa.

Injury to any muscle or tendon aspect of the elbow should be suspected and verified by resisted motion of the involved muscle. Pain can be reproduced from resisting the desired motion of the elbow and eliciting tenderness at the site of insertion of the muscle tendon. Full range of motion may be restricted.

Fractures of the elbow are beyond the scope of this text, but a word of caution in rehabilitating the patient with a fractured elbow is pertinent. Passive exercise, which uses external stress on the elbow to increase or regain range of motion, must be avoided, or at least performed very cautiously. Active exercises performed by the patient are usually more effective and involve less risk of damage. Resisted exercises may also damage tissue about the elbow and must be performed with supervision and caution. Using excessive force to regain range of motion must be avoided.

DISLOCATION

Most dislocations of the elbow are posterior or posterolateral and often may involve a simultaneous fracture of the medial epicondyle. The elbow flexors usually remain fully operational and tend to give dynamic support to the elbow.

Reduction of the dislocation is best done immediately, before the onset of muscular spasm. With the patient lying prone on an examining table, and the elbow hanging from the table, the lateral dislocation is reduced first. Following this reduction, direct traction can be applied. The elbow must not be flexed fully after the reduction, because full flexion may damage the brachial artery. Following reduction, active exercises are performed by the patient. *No* passive exercises should be performed.

NERVE DAMAGE

Because there are several major nerves about the elbow, a dislocation of or direct trauma to the elbow may cause nerve damage. Such damage must always be suspected and evaluated (see Fig. 7-4).

Ulnar Nerve

The ulnar nerve is superficial in the olecranon fossa and therefore is subject to direct trauma. The nerve passes through a groove behind the medial epicondyle, which is covered by a fibrous sheath that forms the cubital tunnel. The nerve then enters the forearm between the two heads of the ulnar flexor muscle.

The ulnar nerve supplies the ulnar flexor muscle of the wrist and the deep flexor muscle of the fingers and enters the hand to supply the muscles of the hypothenar eminence, the third and fourth lumbrical muscles, all the interossei muscles, and the adductor muscle of the thumb (Fig. 7-5). The ulnar nerve also supplies the dermatome area of the hand—the little finger and the ulnar side of the fourth (ring) finger.

The roof of the cubital tunnel is the arcuate ligament, which is taut at 90 degrees of elbow flexion and slack upon elbow extension. The floor of the tunnel is composed of the medial ligament and the tip of the trochlea (Fig. 7-6). The medial ligament bulges in elbow flexion and can compress the nerve. Prolonged periods of extreme elbow flexion must therefore be avoided.

When the arm is abducted, as it is during an intravenous injection, full supination of the forearm removes the possibility of pressure within the tunnel. Pronation in this elbow position is more apt to cause nerve pressure.

Because the sensory fibers of the ulnar nerve are more superficial than the motor fibers, sensory symptoms are more prevalent.

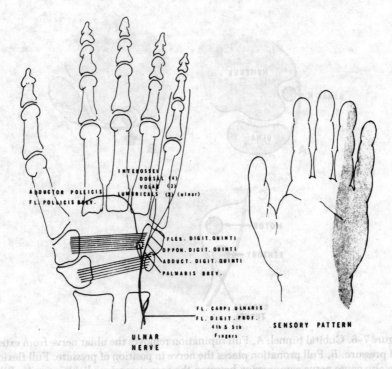

Figure 7-5. The motor and sensory distribution of the ulnar nerve in the hand.

Ulnar nerve paresis resulting from compression is difficult to document; thus, progression of a lesion is difficult to ascertain. An accepted gradation of paresis is offered:

Grade I: Paresthesia and minor hypoesthesia

Grade II: Weakness and wasting of the interossei muscles with incomplete hypoesthesia

Grade III: Paralysis of the interossei muscles with severe atrophy of the hypothenar muscles and of the adductor muscles of the thumb (causing clawing of the ring and little fingers

Treatment. Conservative management of ulnar pressure palsy is usually effective. Such management implies avoidance of pressure upon the cubital canal by use of adequate padding during daily activities, avoidance of excessive elbow flexion, and use of a light splint during sleep.

Surgical transplantation, once frequently performed, is no longer considered valuable. Electrical studies by Payan[1] confirmed that recovery was as rapid with conservative measures as with nerve transposition.

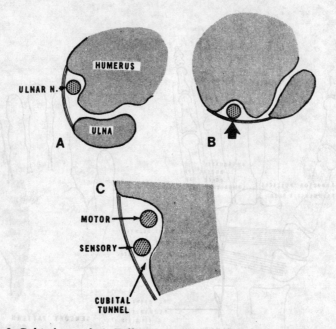

Figure 7-6. Cubital tunnel. *A*, Full supination removes the ulnar nerve from external pressure. *B*, Full pronation places the nerve in position of pressure. Full flexion can also cause nerve compression between the arcuate and medial ligaments of the floor. *C*, Ulnar nerve fibers are divided, with the sensory portion more superficial; thus, pressure often causes sensory changes without motor impairment.

Examination of the patient complaining of paresthesia of the ring and little fingers must rule out cervical root entrapment from diskogenic disease, thoracic outlet neurovascular compression, or pressure of the ulnar nerve at the wrist. Appropriate clinical examination confirmed by appropriate electromyographic (EMG) and conduction studies assists in this differential diagnosis.

Radial Nerve

The radial nerve branches in the region of the elbow, where it is subject to entrapment. In its descent of the lateral aspect of the humerus, it proceeds in front of the lateral condyle of the humerus between the brachial and the brachioradial muscles.

As the radial nerve progresses distally below the elbow joint, it passes under the origin of the short radial extensor muscle (Fig. 7-7). This muscle originates from a fibrous band that stretches from the epicondyle to the deep fascia of the volar surface of the forearm.

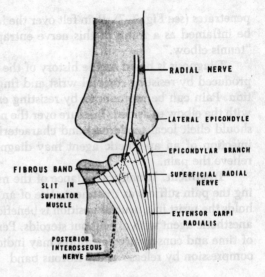

Figure 7–7. Course of the radial nerve. The deep nerve passes under the fibrous band origin of the short radial extensor muscle. At the division just cephalad to the fibrous band, the superficial radial nerve proceeds. After entrance under the band, a small recurrent branch proceeds to the lateral epicondyle.

RADIAL NERVE

LATERAL EPICONDYLE

EPICONDYLAR BRANCH

SUPERFICIAL RADIAL NERVE

EXTENSOR CARPI RADIALIS

FIBROUS BAND

SLIT IN SUPINATOR MUSCLE

POSTERIOR INTEROSSEOUS NERVE

The radial nerve divides at this point with the superficial nerve passing to the outside. The deep branch passes under the fibrous band, where it gives off a small recurrent branch that proceeds to the lateral epicondyle and continues distally to penetrate the supinator muscle through a small slit in the band. This deep branch ultimately becomes the posterior interosseous nerve. During its course, the nerve supplies the muscles that permit dorsiflexion of the wrist and the fingers.

The superficial branch is exposed to direct trauma, which can result in pain or numbness of the lateral aspect of the forearm. Because the sensory dermatomal distribution goes distally to the hand, pain may be referred to the anatomic snuffbox or to the first carpometacarpal joint and may mimic articular conditions in this region.

Fractures or dislocation of the head or the neck of the radius can result in damage to the radial nerve, with pain, tenderness, or motor impairment of the radial nerve distribution. Only the long radial extensor nerve is spared from injury, as this segment usually branches above the elbow joint.

Although direct trauma is frequently the cause of radial nerve mediated pain, these symptoms can occur from trauma imposed by violent muscular contraction of the forearm extensor muscles. This violent contracture may be caused by repeated supination or dorsiflexion against resistance, as occurs in playing tennis, using a screwdriver, or wielding a heavy hammer.

The nerve in these actions becomes trapped by forceful muscular contractions that cause the fibrous band from which the muscles originate to become taut, or by contraction of the small slit through which the nerve penetrates (see Fig. 7–7). Pain felt over the lateral epicondyle, which may be inflamed as a result of this nerve entrapment, may simulate that of "tennis elbow."

Diagnosis is based on the history of the offending activity. Pain is reproduced by resisting forceful wrist and finger extension or wrist supination. Pain can be reproduced by resisting extension of the middle finger with the elbow extended. Pressure over the nerve at the site of entrapment should elicit local tenderness and characteristic radiation of pain. Local injection of an anesthetic agent may diagnostically and therapeutically relieve the pain.

Treatment. Usually, avoidance of the movement or position producing the pain suffices as treatment. Use of an appropriate wrist splint that holds the wrist in a neutral position is beneficial, as is local injection of an anesthetic agent with or without steroids. Persistence of symptoms in spite of time and conservative measures may indicate the need for surgical decompression by release of the fibrous band.

BASEBALL ELBOW

The elbow of a baseball player "wears out" after years of throwing activity. The damage includes elbow flexion contracture, tear or degeneration of the medial ligament, tardy ulnar palsy, and articular cartilaginous degeneration. The flexion contracture results from repeated stretching of the anterior capsule or of the biceps, brachialis, or finger flexor muscles. Periosteal tears are followed by osteogenic buildup. The medial collateral ligament may rupture or undergo attrition.

TENNIS ELBOW

The condition commonly called "tennis elbow" is probably the most frequent sports-related elbow complaint although it also occurs in nonathletes. There are numerous labels attached to this syndrome. In 1936 Cyriax[2] compiled a list of 26 different types of painful lesions affecting the elbow. The tennis elbow syndrome is characterized by an insidious onset of pain that is brought on by wrist extension with pronation or supination and is aggravated by gripping. The pain and tenderness are felt at the area of the lateral epicondyle (Fig. 7–8).

The causative factors are usually repeated minor traumata that strain the wrist extensor musculature at the lateral epicondyle. Repair of the tissue is prevented by repeated trauma from traction, involving a combination of periosteal tearing and microscopic tearing of the fibers of the tendon or the muscle. The most commonly affected muscle is the extensor

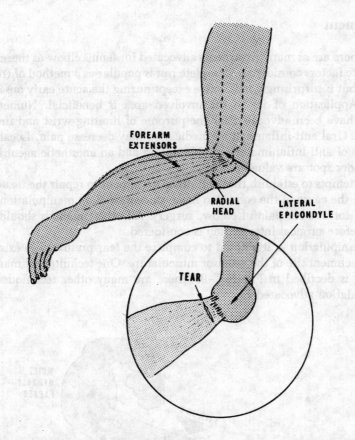

Figure 7–8. Site of tennis elbow (lateral epicondylitis). The forearm muscles that extend the wrist and fingers originate from the lateral epicondyle and extend to a ligament connecting with the head of the radius. Tennis elbow is considered to be a partial tear in the myofascial periosteal tissues of the extensor origin.

carpi radialis brevis, but there may also be involvement of the extensor digitorum communis or the extensor carpi ulnaris. The extensor carpi radialis brevis is the most frequently involved muscle possibly because it is most intimately attached to the joint capsule.

The diagnosis is made by resisting the patient's wrist extension and radial deviation with the elbow fully extended. Pain and tenderness are felt at the lateral epicondyle. To test for involvement of the extensor digitorum communis, the patient simultaneously extends the fingers, and the examiner attempts to flex them during the maneuver.

Treatment

There are as many treatments advocated for tennis elbow as there are etiologic factors considered. Complete rest is popular as a method of treatment, but is surprisingly ineffective except during the acute early onset of pain. Application of ice to the involved area is beneficial. Numerous splints have been advocated for the purpose of limiting wrist and finger flexion. Oral anti-inflammatory medication may decrease pain. Local injections of anti-inflammatory soluble steroid and an anesthetic agent into the tender spot are valuable.

Attempts to establish the "definitive" treatment to repair the tissue or remove the cause of the condition have concentrated on manipulation or, for the intractable painful elbow, surgery. All other methods should be tried before surgical intervention is considered.

Manipulation is attempted to complete the tear presumed to exist at the attachment site of the extensor musculature. One technique of manipulation is decribed in Figure 7-9. There are many other techniques of manipulation advocated.[3]

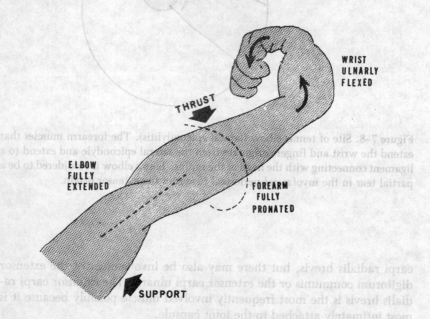

Figure 7-9. Manipulation of tennis elbow. Procedure done with the upper arm supported, the forearm extended and fully pronated, and the wrist and fingers fully flexed in an ulnar direction. A slight brisk thrust to extend the elbow further and pronate the forearm causes a slight snap, which is considered to further tear the attachment and thus give relief after subsequent healing.

When all conservative approaches to treatment have failed, possible surgical procedures include the following:

1. Excision of the damaged portion of the extensor musculature with repair of the defect
2. Excision of the orbicular ligament with release of the extensor mechanism
3. Fasciotomy distal to the extensor tendon
4. Release or lengthening of the extensor carpi radialis brevis
5. Percutaneous tenotomy. This office procedure has recently been advocated.[4]

Other nonoperative procedures have been claimed to provide some relief of symptoms. Whether or not they are effective on their own, their use before and after surgical intervention may help to ensure better recovery from surgery. They include:

1. Strengthening the extensor mechanism by resistive exercises done for long periods of time, especially prior to beginning a program of athletic activities such as tennis.
2. Choice of proper athletic equipment such as rackets of the proper size and circumference.
3. Instruction in proper technique of athletic activities. In tennis, this instruction may include the two-handed backhand technique.

REFERENCES

1. Payan, J: Anterior transposition of the ulnar nerve: An electrophysiological study. J Neurol Neurosurg Psychiatry 33:157–165, 1970.
2. Cyriax, JH: The pathology and treatment of tennis elbow. J Bone Joint Surg 18:921–940, 1936.
3. Kushner, S, and Reid, DC: Manipulation in the treatment of tennis elbow. Journal of Orthopedic and Sports Physical Therapy, March 1986, pp 264–272.
4. Yerger, B, and Turner, T: Percutaneous extensor tenotomy for chronic tennis elbow: An office procedure. Orthopedics 8:1261–1263, 1985.

BIBLIOGRAPHY

Basmajian, JV: Grant's Method of Anatomy. Williams & Wilkins, Baltimore, 1971.
Cailliet, R: Hand Pain and Impairment, ed 3. FA Davis, Philadelphia, 1982.
Cyriax, JH: Textbook of Orthopaedic Medicine, Vol I: Diagnosis of Soft Lesions. Harper & Row, New York, 1954.
Emery, SE, and Gifford, JF: 100 years of tennis elbow. Contemporary Orthopedics, 12 (4):53–58, 1986.

Kopell, HP, and Thompson, WAL: Peripheral Entrapment Neuropathies. Williams & Wilkins, Baltimore, 1963.

Steindler, A: Lectures on the Interpretation of Pain in Orthopedic Practice. Charles C Thomas, Springfield, IL, 1959.

Wadsworth, TG, and Williams, JR: Cubital tunnel external compression syndrome. Br Med J, 3:662–666, 1973.

CHAPTER 8

Wrist and Hand Pain

Hand or wrist pain is both common and frightening to a patient. There are relatively few common causes of pain in the hand. The presence of an infectious lesion is usually obvious, and the history of trauma is usually ascertainable. The other two major causes of hand pain are rheumatologic and neurologic.

As in all musculoskeletal complaints, determining the cause of hand pain requires a thorough understanding of the functional anatomy of the part. In the painful hand, however, innervation of the part plays the major role in diagnosis.

It is not the purpose of this chapter to provide a detailed discussion of fractures of the hand and wrist. These injuries, which are well documented in orthopedic textbooks and journal articles, require proper x-ray verification and precise treatment based on the specific fracture. Therapeutic issues such as open versus closed reduction and the optimum period of immobilization are beyond the scope of this chapter.

Understanding the nerve supply of the hand is mandatory for understanding the basis of hand pain and impairment. The hand is supplied by three major nerves: the median, the ulnar, and the radial nerves (see Figs. 7–5 to 7–7). Nociceptive pain stimuli transmitted via these nerves may result from mechanical trauma or disease.

The functional anatomy has been fully discussed in my book *Hand Pain and Impairment;*[1] thus, only pertinent factors are provided in this chapter.

In 1879, Hilton aptly stated in his treatise *Rest and Pain,* "The same trunk of nerves, the branches of which supply the group of muscles moving any joint, furnish also a distribution of nerves to the skin over these same

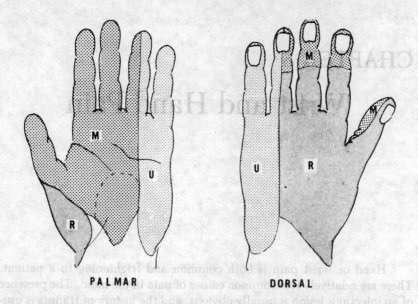

PALMAR   **DORSAL**

Figure 8–1. Sensory map of the hand. Dermatomal areas of the median (M), ulnar (U), and radial (R) nerves. The dorsum of the hand has a great variance, even to not having a radial nerve sensory area.

muscles and their insertions and the interior of the joint receives its nerves from the same source. . . ."[2] This statement applies to *all* extremities, but most specifically to the joints, muscles, and skin of the hand, fingers, and wrist.

The dermatomal mapping of the hand is depicted in Figure 8–1 and comprises all three nerves.

The motor innervation of the hand is supplied essentially by the median nerve (Fig. 8–2) and the ulnar nerve. Because the ulnar nerve is exposed, it is accessible to trauma. Unfortunately, injury to the ulnar nerve is the most crippling in both motor and sensory aspects. Exposure of the ulnar nerve at the elbow has been discussed in Chapter 7.

Pain in the hand may occur from nerve compression in the forearm, but it occurs most often from nerve compression at the wrist. Compression may involve either the median or the ulnar nerve at the wrist with resultant anesthesia, paresthesia, weakness, or atrophy.

The radial nerve supplies merely the skin of the dorsum of the hand, the thumb, and the medial aspect of the hand to the metacarpal joints. The dermatomal region of the radial nerve is depicted in Figure 8–3.

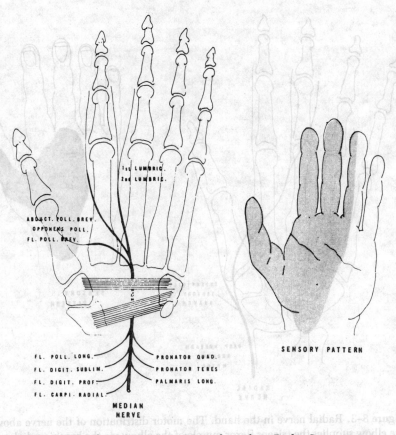

1st LUMBRIC.
2nd LUMBRIC.

ABDUCT. POLL. BREV.
OPPONENS POLL.
FL. POLL. BREV.

FL. POLL. LONG.
FL. DIGIT. SUBLIM.
FL. DIGIT. PROF.
FL. CARPI. RADIAL.

PRONATOR QUAD.
PRONATOR TERES
PALMARIS LONG.

MEDIAN
NERVE

SENSORY PATTERN

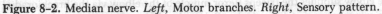

Figure 8–2. Median nerve. *Left,* Motor branches. *Right,* Sensory pattern.

CARPAL TUNNEL SYNDROME

The carpal tunnel is a narrow fibro-osseous tunnel through which traverse ten structures: the flexor pollicis longus tendon, four flexor digitorum superficialis tendons, four flexor digitorum profundus tendons, and the median nerve (Fig. 8–4). The distal volar skin crease of the wrist is the visible proximal border of the canal. The canal (within the tunnel) extends distally approximately 3 cm.

The roof of the tunnel is the transverse carpal ligament, which comprises two bands: one from the hook of the hamate bone extending to the tubercle of the trapezium bone and a proximal band extending from the tubercle of the navicular (scaphoid) bone to the pisiform bone (Fig. 8–5). The floor of the tunnel is composed of the carpal bones of the hand.

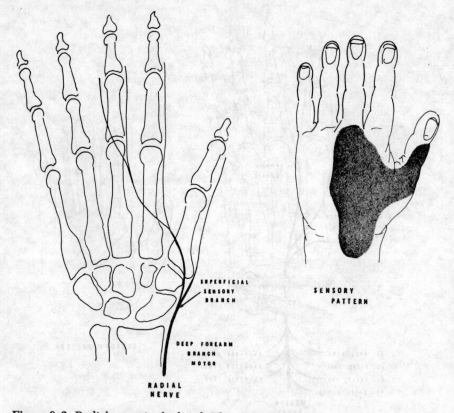

Figure 8–3. Radial nerve in the hand. The motor distribution of the nerve above the elbow supplies the triceps flexor muscle of the elbow via the brachioradial and the extensor carpi radial muscles. Below the elbow, the nerve supplies ulnar wrist extensors, extensors of the fingers, and extensors of the distal phalanx of the thumb and index finger.

Symptoms

The symptoms complained of by the patient are pain described as "burning," and numbness or paresthesia, that is, tingling or "falling asleep" *of the hand and of the fingers supplied by the median nerve* (see Fig. 8–2). These symptoms may awaken the patient from sleep, and the patient may claim that relief is obtained by shaking or rubbing the hands. Symptoms are increased by prolonged gripping of a steering wheel or of a book or by prolonged knitting.

The symptoms are usually unilateral, but some patients complain of symptoms in both hands. This syndrome is more prevalent among women.

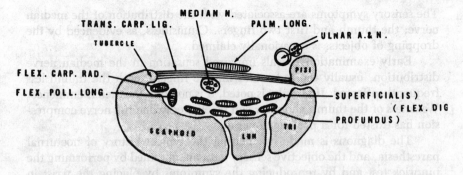

Figure 8–4. Contents of the carpal tunnel. The tunnel described in Figure 8–5 contains the deep and superficial long finger flexor tendons, the tendons of the long flexor muscles of the thumb and of the ulnar flexor muscle of the wrist, and the median nerve.

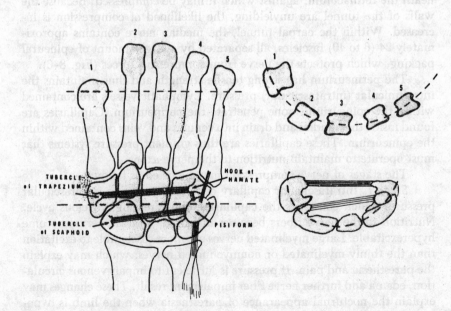

Figure 8–5. Transverse carpal ligaments. These ligaments bridge the arch of the carpal rows and form a tunnel. The proximal band extends from the tubercle of the navicular bone to the pisiform bone and the distal band from the tubercle of the trapezium bond to the hook of the hamate bone.

The sensory symptoms are associated with the distribution of the median nerve: the thumb and first two fingers. Clumsiness, as evidenced by the dropping of objects, is occasionally claimed.

Early examination reveals impaired sensation in the median nerve distribution, usually the index and middle fingers, with the thumb less frequently affected. If atrophy is noted, it is noted in the thenar eminence. Weakness of the thumb abductor can be elicited when the nerve compression has existed for a prolonged period.

The diagnosis is made by eliciting the typical history of nocturnal paresthesia, and the objective sensory data are obtained by performing the pinprick test and by reproducing the symptoms by placing the wrist in forced flexion and maintaining this position for 60 to 90 seconds.

Confirmation of median nerve compression within the carpal tunnel may be accomplished by performing an electromyographic study of nerve conduction time. The nerve velocity is normal from the elbow to the wrist, and then is delayed across the transverse carpal ligament into the hand and fingers.

The mechanism and structural changes within the nerve have recently been summarized by Sunderland.[3] The nerve is located directly beneath the retinaculum, against which it may be compressed. Because the walls of the tunnel are unyielding, the likelihood of compression is increased. Within the carpal tunnel, the median nerve contains approximately 24 (6 to 40) fascicles, all separated by a large amount of epineural packing, which protects the nerve from compressive forces (Fig. 8–6).

The perineurium has strong tensile strength and thus maintains the intrafunicular (intrafascicular) pressure. Lymphatic vessels are contained within the epineurium; none penetrate the perineurium. Capillaries are found inside the bundles and drain into venules and veins contained within the epineurium. These capillaries are thus gradient pressure systems that must operate to maintain nutrition to the nerve axons.

The stages of nerve compression may be listed as follows:

Stage I. Intrafascicular capillary distension increases intrafascicular pressure, which constricts the capillaries, thus creating a vicious cycle. Nutrition to the nerve fibers becomes impaired, and the nerves become hyperexcitable. Large myelinated nerves are more susceptible to excitation than the thinly myelinated or nonmyelinated nerves, which may explain the paresthesia and pain. If pressure is sufficient to impair venous circulation, edema and further nerve fiber impairment result. These changes may explain the nocturnal appearance of paresthesia when the limb is hypotonic and dependent and the decrease of the paresthesia when the arm is elevated and exercised. Blood pressure cuff compression aggravates the paresthesia on this basis.

Stage II. As capillary compression occurs, anoxia develops, which damages the capillary endothelium. Protein leaks into the tissues, creating

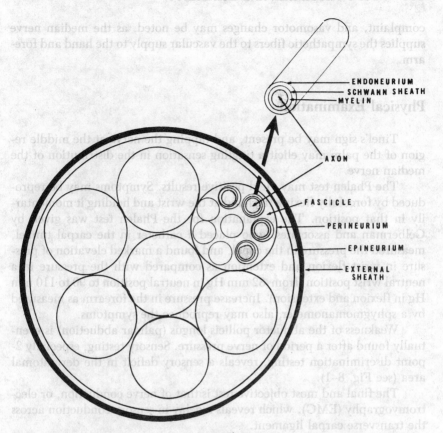

Figure 8–6. Schematic representation of peripheral nerve. Cross section shows that nerve is composed of many axons grouped into a fascicle. Each axon is surrounded by myelin enclosed within a sheath of Schwann. This sheath is coated with endoneurium, which is composed of longitudinal collagen strips. Perineurium binds the fascicles, which are in turn bound together by epineurium. The entire nerve is covered by an external sheath.

more edema. Protein cannot escape the perineurium, so fluid accumulates within the endoneural space; this fluid accumulation interferes with axon nutrition and metabolism. Fibroblasts proliferate in this ischemic atmosphere, and scarring causes formation of a constrictive connective tissue. At this later stage, the nerve lesions are irreversible, which explains the failure of long-standing sensory and motor defects to improve after decompression.

Weakness and diminished use of the hand, especially the thumb, may be a complaint, but this weakness may be the examiner's objective finding when there has been no concern expressed by the patient. Aching may be a complaint, and vasomotor changes may be noted, as the median nerve supplies the sympathetic fibers to the vascular supply to the hand and forearm.

Physical Examination

Tinel's sign may be present, and tapping the nerve at the middle region of the palm may elicit a tingling sensation in the distribution of the median nerve.

The Phalen test may yield positive results. Symptoms may be reproduced by forcefully flexing the hand at the wrist and holding it momentarily in that position. The explanation for the Phalen test was given by Gelberman and associates, who placed a catheter in the carpal tunnel, measured the pressure in the canal, and found a marked elevation of pressure in wrist flexion and extension as compared with the pressure in a neutral wrist position (from 32 mm Hg in neutral position to 90 to 110 mm Hg in flexion and extension). Increase pressure in the forearm as measured by a sphygmomanometer, also may reproduce the symptoms.

Weakness of the abductor pollicis longus (palmar abduction) is eventually found after a period of nerve pressure. Sensory testing, especially 2-point discrimination testing, reveals a sensory deficit in the dermatomal area (see Fig. 8–1).

The final and most objective test is that of nerve conduction, or electromyography (EMG), which reveals a delay in sensory conduction across the transverse carpal ligament.

Differential Diagnosis

Other entrapment syndromes of the median nerve must be considered as possibly causing the symptoms.

Anomalous Median Nerve Entrapment. The ligament of Struthers at the distal humerus connecting to an anomalous bony spur may entrap the median nerve. X-ray studies may reveal this spur, and a conduction time study may reveal the delay at this site.

Pronator Teres Syndrome. In the antecubital area, the median nerve enters the forearm between the two heads of the pronator teres muscle, then passes under the edge of the flexor digitorum sublimis muscle. Entrapment of the nerve at this site is suspected when the patient has tenderness of the forearm and when the paresthesia occurs from activities involving excessive forearm pronation. Tinel's sign is present at the pronator teres site, and symptoms are reproduced by forcefully pronating the forearm while the elbow is slowly extended.

Anterior Interosseous Nerve Syndrome. The anterior interosseous nerve arises from the median nerve as the median nerve passes between the two heads of the pronator teres muscle. There are no skin sensory nerve fibers in this nerve, but the motor fibers innervate the flexor pollicis longus, flexor digitorum profundus, and the pronator quadratus. Clinically, the patient cannot flex the distal phalanx of the thumb and index finger, and wrist pronation is weak. Electromyography reveals abnormalities of the involved muscles.

Treatment

Treatment of carpal tunnel syndrome requires splinting the wrist in a neutral position day and night for several weeks. Splinting the wrist only at night when the symptoms appear is ineffective. Administration of diuretics has been suggested to relieve the swelling within the canal. Some physicians advise injecting a soluble steroid into the carpal tunnel.

When the symptoms and objective findings such as weakness and hypalgesia persist despite appropriate conservative treatment, surgical decompression is indicated. Proper surgical decompression, that is, release of both bands of the transverse carpal tunnel, is effective and reasonably simple. Prolonged median nerve compression leads to permanent sensory loss and some motor loss.

Rheumatoid arthritis, with its accompanying tendinitis, frequently causes nerve compression within the carpal tunnel. Surgical decompression should be considered early when pharmaceutical treatment fails to afford subjective relief. Steroids injected under the carpal ligament have some diagnostic value but produce limited therapeutic benefit.

ULNAR NERVE COMPRESSION

At the wrist, the ulnar nerve passes into the hand at Guyon's canal (Fig. 8-7), which is a shallow trough between the pisiform bone and the hook of the hamate bone. Its floor is a thin layer of ligament and muscle. The roof is the volar carpal ligament and the long palmar muscle.

After the nerve emerges from the tunnel, it divides into two ulnar branches, which convey sensation from the side of the palm and the fourth and fifth fingers. The deep branch of the nerve supplies the muscles of the hypothenar eminence, the third and fourth lumbrical muscles, all of the interossei muscles, adductor muscles of the thumb, and the deep head of the short flexor muscle of the thumb.

A lesion at the wrist may cause (1) motor and sensory deficit if the trunk is involved, (2) predominantly sensory loss if the superficial nerve is

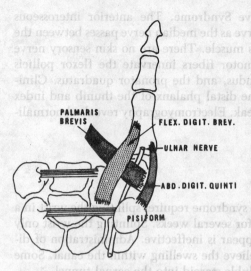

Figure 8-7. Guyon's canal. The ulnar nerve enters the hand at the wrist via a shallow trough between the pisiform bone and the hook of the hamate bone. It is covered by the volar carpal ligament and the long palmar muscle.

involved, or (3) primary motor deficit if the lesion is deep.

The most common cause of ulnar nerve involvement is pressure such as that produced by habitually leaning on the elbow, by trauma affecting baseball pitchers, or by pressure occurring after an operation or procedure that requires anesthetizing a patient and allowing a prolonged direct pressure upon the elbow. Use of narcotics frequently causes this positional ulnar pressure. Diabetes and alcoholism are also frequent causes.

Symptoms

There is sensory involvement of the volar and dorsal aspects of the small and ring fingers. There may be weakness of the intrinsic muscles of the hand and weakness of the flexor carpi ulnaris and flexor profundus muscles of the little and ring fingers. Percussion test results are often positive. X-ray studies should be performed to rule out bone or joint disorders.

Treatment

Protection of elbow pressure is a valid objective of treatment. Use of a sponge elbow pad that has a hole cut out over the ulnar nerve region is effective. The daily habits of sitting and working must be modified to minimize elbow pressure. Oral anti-inflammatory drugs merit consideration, as does local infiltration into the canal of soluble steroids and an anesthetic agent.

After failure of conservative measures, surgical decompression and/or nerve transposition can be considered, although the long-range results of these procedures are sometimes questioned. The presence of objective nerve impairment justifies the procedure.

Ulnar nerve symptoms must always be considered as possibly originating at the elbow or from a nerve root of the cervical brachial plexus. These conditions have been discussed in Chapters 6 and 7.

Treatment may merely require steroid and analgesic infiltration into the canal. Should symptoms persist and lead to severe neurologic deficit, surgical decompression should be considered.

Tendon Sheath Pathology

Pain in the wrist and hand can originate within the tendon sheaths from injury, infection, or severance. These tendons may be either flexors or extensors. The specific site of pain or the painful motion depends on the specific tendon involved. Excessive repetitive movements or unphysiologic stress upon the tendon may inflame the sheath with resultant painful limited motion. The tendons usually swell, and crepitation can be elicited during motion. The tendons of the wrist most commonly involved are the dorsal extensors of the wrist and the long abductor and short extensor of the thumb. These structures form the anatomic site termed the snuffbox (Fig. 8–8). Tendinitis of the snuffbox is Quervain's disease.

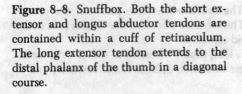
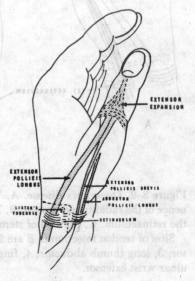

Figure 8–8. Snuffbox. Both the short extensor and longus abductor tendons are contained within a cuff of retinaculum. The long extensor tendon extends to the distal phalanx of the thumb in a diagonal course.

QUERVAIN'S DISEASE

Originally described in 1895 by the Swiss surgeon Fritz de Quervain, stenosing tenosynovitis of the thumb abductor at the radiostyloid process is very common. The long abductor tendon and the short extensor tendon move in a common sheath that passes in a bony groove over the radiostyloid process (Fig. 8–9). Distal to the process, the tendons form a sharp angle of approximately 105 degrees that encourages friction with resultant synovitis. During thumb-pinching activities, the long abductor muscle stabilizes the thumb, which causes friction.

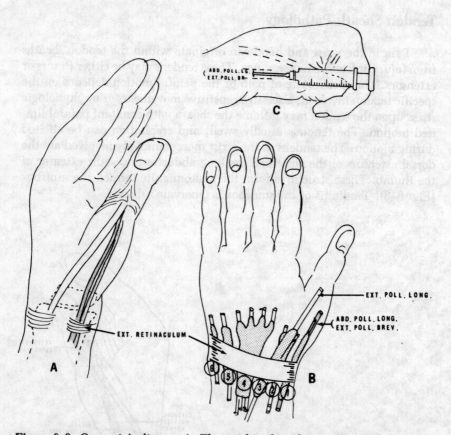

Figure 8–9. Quervain's disease. *A*, The combined tendons pass over the prominence of the radiostyloid process. *B*, The extensor tendons of the fingers pass under the retinaculum. *C*, Method of steroid injection.

Sites of tendon injection in *B* are 1, long thumb extensor; 2, radial wrist extensor; 3, long thumb abductor; 4, finger extensors; 5, little finger extensor; and 6, ulnar wrist extensor.

Symptoms include an aching discomfort over the styloid process that is aggravated by movements of the wrist and thumb. Abduction of the thumb against resistance can reproduce the symptoms, and there may be tenderness over the tendon. A diagnostic procedure consists of reproducing the symptoms by flexing the thumb and cupping it under the fingers, and then flexing the wrist in an ulnar direction. This maneuver stretches the thumb tendons and causes pain.

The pathologic process is a combination of edema within the sheath and increased vascularity of the sheath. The sheath constricts the tendons and restricts movement. Fine adhesions have been described in cases of persistent synovitis.

Treatment demands immobilization of the thumb and wrist in a padded mold cast. Injection of steroids into the sheath is beneficial (see Fig. 8–9B and 8–9C).

If symptoms persist after 3 to 4 weeks of immobilization and a series of injections, surgical decompression is indicated. It must be ascertained at surgery that both tendons (long thumg abductor and short thumb extensor) are within the sheath, or surgical decompression will be ineffectual.

TEAR OF THE LONG EXTENSOR TENDON OF THE THUMB

The long extensor tendon of the thumb can rupture because of the sharp angulation of the tendon about the tubercle of Lister and because of friction there. Rheumatoid arthritis or damage from a Colles's fracture may hasten rupture. This rupture can be diagnosed by the patient's inability to extend the distal phalanx of the thumb. Conservative treatment for this condition is ineffectual, and surgical repair is indicated.

TRIGGER FINGER

A sudden snapping sensation of a finger during flexion and re-extension may be noted, and actual locking of the finger may result. Once locked, the finger cannot further flex or extend. This condition can occur in the thumb or in any of the fingers, and it usually occurs in the flexor tendons.

A nodule forms on the tendon within its thickened synovium-lined sheath. When the nodule becomes too thick, obstruction occurs. Figure 8–10 depicts the more common snapping third and fourth fingers, and Figure 8–11 depicts the snapping thumb.

Usually, the nodules can be palpated, and the condition of snapping and locking can be demonstrated by the patient.

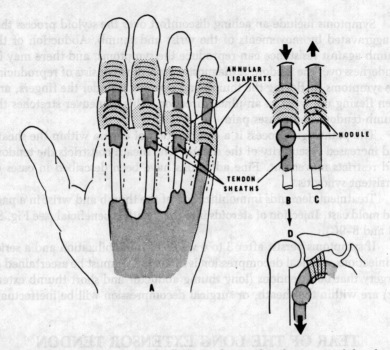

Figure 8–10. Trigger finger. *A*, The flexor tendons within their synovial sheath pass under the annular ligaments at the metacarpal heads. *B*, Nodule proximal to the ligament prevents flexion. *C*, Nodule is trapped. *D*, Re-extension is prevented.

Cortisone injection within the sheath may result in complete and permanent recovery. Should a series of injections fail, surgical intervention should relieve the problem. The annular band (see Fig. 8–10) is slit to permit the nodule to pass, whereas excision of the nodules may cause a larger nodule to form.

SPRAINS AND DISLOCATIONS

Sprains may involve momentary subluxation that reduces spontaneously. Because the subluxation reduces, normal x-ray pictures are produced, and the soft tissue injury escapes detection. The capsule and collateral ligaments may be torn, and actual articular dislocation may occur.

Joint limitation, pain, and swelling may be evident on examination. The injured joint's passive range of motion, when compared with that of the normal contralateral joint, may reveal excessive mobility, but joint restriction is usually evident.

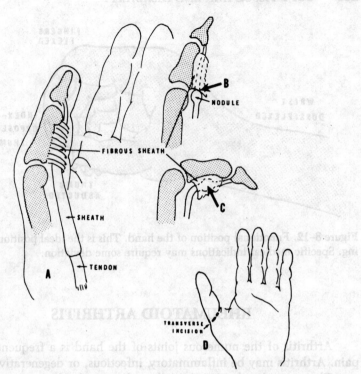

Figure 8–11. Snapping thumb flexor. *A*, Flexor tendon under its sheath passing through the fibrous canal. *B*, Nodule that prevents flexion. *C*, Nodule trapped within sheath, which prevents re-extension. *D*, Site and direction of decompressing the tenosynovitis.

When a sprain is suspected, immobilization of the finger in a slightly flexed functional position for 2 to 3 weeks usually permits the soft tissue injury to heal (Fig. 8–12). The immobilization should be followed by progressive active exercise.

Subluxation of the metacarpophalangeal joint may tear the palmar plate (Fig. 8–13). Palmar plates are fibrocartilaginous plates that reinforce the joint capsule to prevent hyperextension and prevent excessive friction of the flexor tendons. The distal portion of the plate is cartilaginous, and the proximal portion is membranous. The distal portion is firmly attached to the phalanx, but the proximal portion is loosely attached to the metacarpal bone.

The palmar plate, because of its membranous portion, retracts when permitted to remain in a shortened position and thus forms contracture. Prolonged immobilization of a finger in flexion results in a fixed flexion contracture.

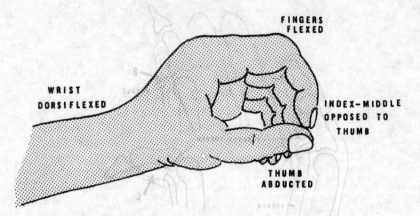

Figure 8–12. Functional position of the hand. This is the ideal position for splinting. Specific medical indications may require some deviation.

RHEUMATOID ARTHRITIS

Arthritis of the numerous joints of the hand is a frequent cause of pain. Arthritis may be inflammatory, infectious, or degenerative.

Rheumatoid arthritis is initially a disease of soft tissue, a disease of synovium. Patients with rheumatoid arthritis have a high incidence of involvement of tendons and their sheaths. As tenosynovitis proceeds, granulomatous synovium invades the tendon, causing it to weaken, lengthen, and frequently rupture. Simultaneously, the synovitis invades all the periarticular tissues, the capsule ligaments, and ultimately the cartilage. Full discussion of the pathology and manifestations of rheumatoid arthritis in regard to each specific joint or joints is beyond the scope of this book.

The joints of the hand most frequently involved are the proximal interphalangeal joint, the metacarpophalangeal joints, and the wrist joint. Loss of function of the proximal interphalangeal joint and the metacarpophalangeal joints is far more disabling than loss of function of the distal phalangeal joints. Impaired thumb motion causes a major functional handicap. If the carpometacarpal and metacarpophalangeal joints of the thumb are involved, rotation of the thumb is restricted, and tip-to-tip opposition is lost.

Treatment

Treatment of an acute condition of the rheumatoid hand focuses on reduction of the inflammation, swelling, and pain. Obviously, treatment

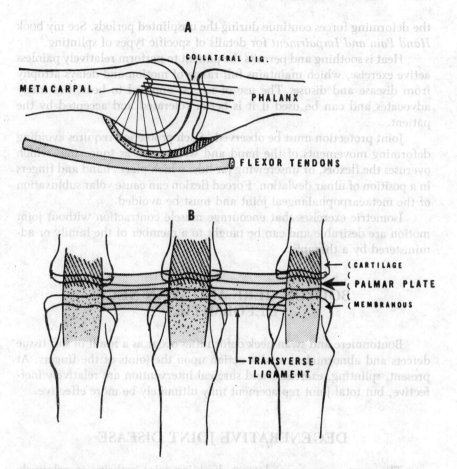

Figure 8–13. Palmar plate. There is a fibrocartilaginous plate on the palmar surface of the joints that reinforces the capsule. *A*, The proximal portion is membranous and is loosely attached to the metacarpal. The distal portion is cartilaginous and is firmly attached to the phalanx. *B*, The plates are connected to the deep transverse ligament. This ligament also prevents lateral motion of all fingers but the thumb.

must be directed toward the systemic disease, but specific treatment of the hand must be seriously considered.

Rest of the hand is mandatory but is extremely difficult to accomplish. Movement during the acute phase is potentially detrimental to hand structure, yet activities of daily living require some use of the hands. Splints to mobilize the hand and fingers are cumbersome, difficult to apply and maintain, and frequently tax the patient's tolerance and cooperation. Splints can be applied for varying periods during the day and at night, but the deforming forces continue during the unsplinted periods. See my book *Hand Pain and Impairment* for details of specific types of splinting.[1]

Heat is soothing and permits the patient to perform relatively painless active exercise, which maintains full range of motion and delays atrophy from disease and disuse. The use of ice, as opposed to heat, has many advocates and can be used if it is better tolerated and accepted by the patient.

Joint protection must be observed.[4] Such protection requires avoiding deforming movements of the hand and wrist, such as squeezing, which overuses the flexors, or unscrewing jar lids, which places hand and fingers in a position of ulnar deviation. Forced flexion can cause volar subluxation of the metacarpophalangeal joint and must be avoided.

Isometric exercises that encourage muscle contraction without joint motion are desirable and can be taught to a member of the family or administered by a therapist.

BOUTONNIERE AND SWAN NECK DEFORMITIES

Boutonniere and swan neck deformities occur as a result of soft tissue defects and abnormal muscular action upon the joints of the fingers. At present, splinting, exercises, and surgical intervention are relatively ineffective, but total joint replacement may ultimately be more effective.

DEGENERATIVE JOINT DISEASE

The most common and most disabling joint arthritis in relatively young people is degenerative disease of the thumb carpometacarpal joint (Fig. 8–14). In this condition, there are tenderness, stiffness, pain, and crepitation on movement. Grip becomes impaired, and fine movements requiring tip-to-tip activity of the thumb are restricted. This condition is usually bilateral and is more prevalent among women.

Relief is afforded by resting the part. Often, a leather or plastic molded splint can be used for this purpose. Intra-articular injections of steroids provide excellent but temporary relief (Fig. 8–15). Surgical fusion, resection of the trapezium, or silicone implant may be valuable.

VASCULAR IMPAIRMENT OF THE HAND

Raynaud's phenomenon is a condition in which spasm of small blood vessels occurs, especially in the fingers. The patient may observe intermit-

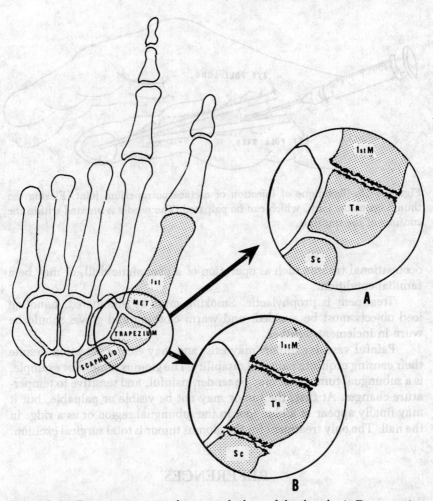

Figure 8–14. Degenerative joint disease at the base of the thumb. *A,* Degenerative arthritis of the first carpometacarpal joint responds well to treatment by splinting, injection, or surgery (excision of trapezium fusion or implant). *B,* When there are arthritic changes in the trapezioscaphoid joint, arthrodesis or implant may not be of value.

tent attacks of sudden pallor to actual blanching of the fingers. Pain may result. The acute episode of pallor may be followed by cyanosis.

The condition is attributed to arterial spasm, a manifestation of vasomotor instability. It is usually triggered by cold or emotional stress and is most prevalent in middle-aged females.

A Raynaud-like phenomenon of the hands may result from neurovascular compression of the cervical dorsal outlet, may result from repeated

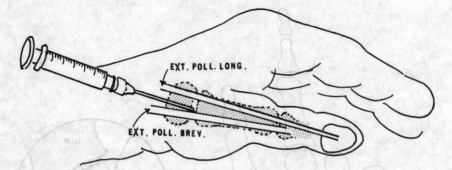

Figure 8–15. Technique of injection of metacarpotrapezium joint. Flexing the thumb opens the joint, which can be palpated. The needle is inserted within the confines of the snuffbox.

occupational trauma such as operation of a pneumatic drill, or may be a familial condition.

Treatment is prophylactic. Smoking must be eliminated, handling iced objects must be avoided, and warm clothing and gloves should be worn in inclement weather.

Painful vascular tumors may exist and may escape detection despite their causing exquisite pain and disability. The glomus tumor, for example, is a subungual tumor that may be tender, painful, and sensitive to temperature changes. At first, the tumor may not be visible or palpable, but it may finally appear as a blue spot in the subungual region or as a ridge in the nail. The only treatment for the glomus tumor is total surgical excision.

REFERENCES

1. Cailliet, R: Hand Pain and Impairment, ed 3. FA Davis, Philadelphia, 1982.
2. Hilton, J: Rest and Pain. Wm Wood & Co, New York, 1879.
3. Sunderland, S: The nerve lesion of the carpal tunnel syndrome. J Neurol Neurosurg Psychiatry 39:615–626, 1976.
4. Melvin, JL: Joint protection training and energy conservation. In Rheumatic Disease: Occupational Therapy and Rehabilitation. FA Davis, Philadelphia, 1977.

BIBLIOGRAPHY

Allen, EV, and Brown, GE: Raynaud's disease: A clinical study of 147 cases. JAMA 99:1472, 1932.
Boyes, JH (Ed): Bunnell's Surgery of the Hand, ed 5. JB Lippincott, Philadelphia, 1970.
Carroll, RE, and Berman, AT: Glomus tumors of the hand. J Bone Joint Surg [Am] 54:697, 1972.

Carstam, N, Eiken, O, and Andren, L: Osteoarthritis of the trapezio-scaphoid joint. Acta Orthop Scand 39:354, 1968.

Chase, RA: Surgery of the Hand. N Engl J Med 287:1174, 1972.

Flatt, AE: The Care of the Rheumatoid Hand. CV Mosby, St. Louis, 1963.

Garland, H, Sumner, D, and Clark, JMP: Carpal-tunnel syndrome: With particular reference to surgical treatment. Br. Med J, March 2, 1963, pp 581–584.

Gelberman, RH, Hergenroeder, PT, Hargens, AR, et al: The carpal tunnel syndrome. J Bone Joint Surg 63-A:380–383, 1981.

Goldner, JL: Symposium: Upper extremity nerve entrapment syndrome. Contemporary Orthopedics, 6:89–112, June 1983.

Kapell, HP, and Thompson, WAL: Peripheral Entrapment Neuropathies. Williams & Wilkins, Baltimore, 1963.

Otto, N, and Wehbe, MA: Steroid injections for synovitis in the hand. Orthopedic Review 15:45–48, May 1986.

Strickland, JW: Flexor tendon injuries, part 1. Anatomy, physiology, biomechanics, healing and adhesion formation around a repaired tendon: A review series. Orthopaedic Review 15:21–34, October 1986.

Wertsch, JJ, and Melvin, J: Median nerve anatomy and entrapment syndromes: A review. Arch Phys Med Rehabil, 63:623–627, 1982.

Wynn Parry, CB: Rehabilitation of the Hand. Butterworths, London, 1966.

Zancolli, E: Structural and Dynamic Basis of Hand Surgery. JB Lippincott, Philadelphia, 1968.

CHAPTER 9
Hip Joint Pain

The hip joint in man is used predominantly for weight bearing but is intrinsically involved in ambulation. It is well constructed for both of these functions, and despite the numerous and varied types of trauma imposed on it in everyday activities, it is infrequently impaired.

FUNCTIONAL ANATOMY

The hip joint is a congruous joint, that is, a joint in which the convex surface of an articular surface is symmetric to the opposing concave articular surface. In this case, the surfaces of the femoral head are opposed to the concave acetabulum.

The head of the femur is spherical and points medially upward and forward (Fig. 9–1). It articulates with the acetabulum, but because of the angle of its neck, the anterior portion of the head of the femur is not engaged in the socket in the neutral leg position.

The acetabulum is horseshoe-shaped and is covered peripherally with cartilage. The center of the horseshoe is not covered with cartilage. The open lower portion of the acetabulum is completed into a total ring by the transverse acetabular ligament (Fig. 9–2). The acetabulum is deepened by a complete ring of fibrocartilage, which is termed the **labrum**. The head of the femur is held firmly within the acetabulum by a thick capsule (see Fig. 9–2). The fibers of the capsule are oblique and become taut when the hip is extended and rotated. There are portions of the capsule that are thickened to form ligaments. These ligaments are termed according to their specific

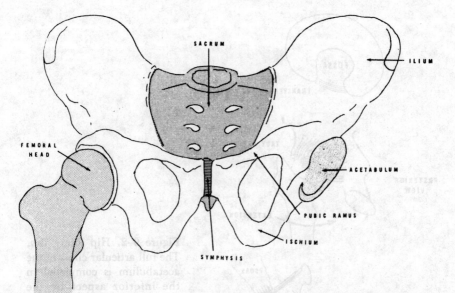

Figure 9-1. Anterior view of bony pelvis. The pelvis is pictured anteriorly with the left femur omitted to reveal the acetabulum.

site of origin: the iliofemoral, the pubofemoral, and the ischiofemoral ligaments.

In the erect stance, the center of gravity passes behind the center of the hip joint. The pelvis normally is rotated forward to seat the hip joint and form the sacral base of the lumbar spine. The anterior portion of the capsule is thickened to form the iliofemoral ligament, which serves to resist excessive backward rotation.

In a toe-out stance, the head of the femur is directed forward out of the socket. The iliofemoral ligament moves laterally, thus depriving an anterior portion of the hip joint of ligamentous support. This exposed area is covered by the tendon of the psoas muscle. The portion of the joint that does not have cartilage is lined with synovial membrane, which extends to completely encircle the neck of the femur.

Another ligament connects the head of the femur to the center of the acetabulum within the centrum of the hip joint. This ligament is a hollow tube of synovial membrane that transmits blood vessels—the branches of the medial circumflex and obturator arteries—to the head of the femur.

Blood is supplied to the hip joint by the medial femoral circumflex, lateral femoral circumflex, and obturator arteries, and by branches of the gluteal artery.

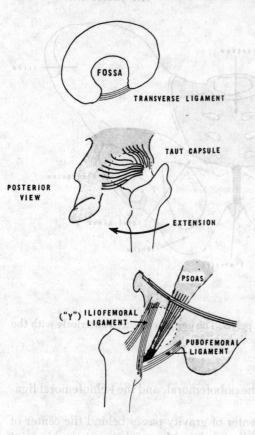

Figure 9-2. Hip joint. *Top*, The full articular circle of the acetabulum is completed in the inferior aspect by the transverse acetabular ligament. *Center*, The oblique capsular fibers become taut as the hip extends. *Bottom*, The anterior capsule is reinforced by the iliofemoral, pubofemoral, and ischiofemoral ligaments and the psoas tendon.

HIP JOINT MOTIONS

The hip joint motions are depicted in Figures 9-3 and 9-4. Flexion is limited by the hamstring muscles when the knee is extended and by contact with the abdominal wall when the knee is flexed. Extension is limited by the ligamentous thickening of the fibrous capsule. With the hip extended, the capsular fibers limit internal and external rotation. Abduction is limited by the abductor muscle group and adduction by the tensor muscle of the fascia lata and the abductor muscle group.

In a relaxed stance, the hip is fully extended. Hyperextension of the hip is achieved by rotation of the pelvis and extension of the lumbar spine into further lordosis.

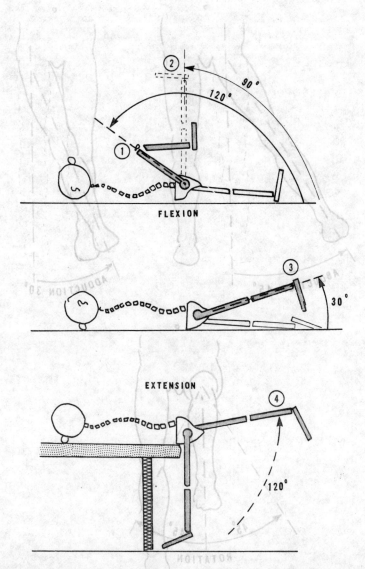

Figure 9–3. Hip joint range of motion. 1, Hip flexion with knee flexed from 0 to 120 degrees limited by contact of the thigh with the abdominal wall. 2, Hip flexion with knee extended 90 degrees is limited by the hamstring muscles. 3, Hip extension 30 degrees with patient prone. 4, With opposite leg flexed to 90 degrees when patient lies prone on examination table, hip flexion of tested leg should extend to 90 to 120 degrees. This position permits hip flexion contracture to be measured clinically.

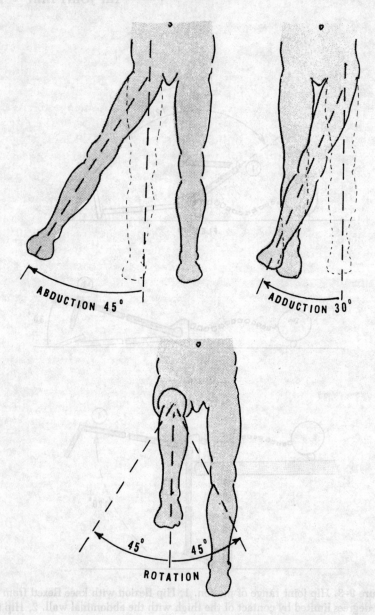

Figure 9–4. Hip joint range of motion. Hip abduction is normally 45 degrees from 0 degrees, measured with the leg fully extended, and adduction is 30 degrees. Rotation of hip measured with knee flexed is normally 45 degrees of internal rotation and 45 degrees of external rotation.

Walking

Walking exerts repeated stretching of the hip capsule, ligaments, fascia, and muscles of the flexor aspect of the hip joint. This stretching occurs because every alternate step of the gait requires that the leg be fully extended (Fig. 9–5). The natural tendency of fibrous connective tissue to shorten causes hip flexion to become contracted and resist full extension. Habitual sitting and lack of exercising predisposes hip flexors to contracture with resultant impairment of gait and adverse effect upon the lumbar spine. As flexion contracture occurs, the iliotibial band similarly contracts, thus producing abduction and external rotation of the hip.

In normal walking, the determinants of gait that level out the center of gravity are pelvic rotation (Fig. 9–6), pelvic tilt (Fig. 9–7), lateral displacement of the pelvis, and knee flexion in the stance phase (Fig. 9–8). All these determinants require adequate range of motion of the hip.

In normal gait, the femur rotates upon the pelvis, and the tibia rotates upon the femur. These movements also require normal range of motion of the hip (Fig. 9–9). Loss of hip motion impairs gait and decreases efficiency, grace, and conservation of energy.

The average normal gait requires 60 percent of its timing in the stance phase and 40 percent in the swing phase (Fig. 9–10). Patients with hip disease spend a disproportionate amount of time in the stance phase of the involved extremity. Patients with diseased hips also excessively rotate laterally over the affected hip during the stance phase. These variations change the gait velocity, stride length, and cadence in the individual's attempt to minimize pain and improve stability. The sum contribution of these changes to the pathologic condition of the head of the femur or the acetabulum has not been determined.

Figure 9–5. Undulant course of pelvic center in gait without determinants. The marked vertical undulations of the center of gravity occur when gait is performed with stiff knees and no lateral motion of the pelvis.

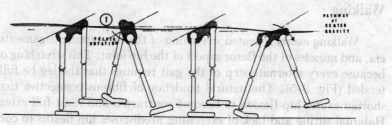

Figure 9-6. Pelvic rotation: determinant of gait. The pelvis rotates forward with the swinging leg aimed to decrease the angle of the leg to the floor measured at the hip joint. This decreases the vertical undulation of the center of gravity.

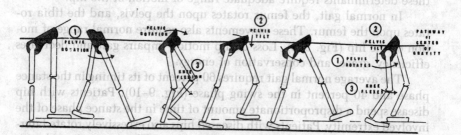

Figure 9-7. Pelvic tilt: determinant of gait. As the left leg swings through, the pelvis drops on the left (2). The left hip and knee flex (3). The last figure depicts the right leg swinging through with the right pelvis dipping and the right hip and knee flexing. This movement decreases the center of gravity undulation, but requires full range of motion of the hip.

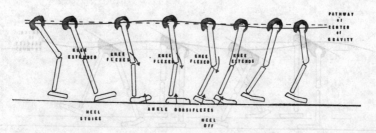

Figure 9-8. Knee flexion during stance phase: determinant of gait. The knee is fully extended at heel strike. As the body passes over the center of gravity, the knee flexes to decrease the vertical amplitude of the pathway of the center of gravity. The knee re-extends at the end of the stance phase: the "heel off."

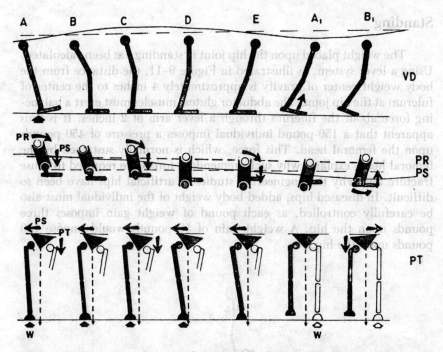

Figure 9-9. Composite schematic determinants of gait. VD indicates the vertical displacement of the pelvis from the side view. PR is pelvic rotation viewed from above as the left leg swings through. PT depicts pelvic tilting. The bottom figure shows the weight bearing leg (W) going into a Trendelenburg position as the hip adducts. PS indicates the pelvic shift. All these determinants require good hip motion.

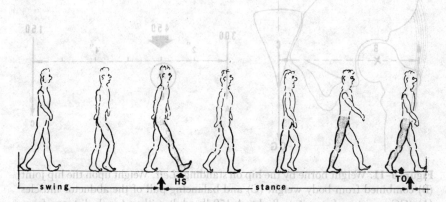

Figure 9-10. Gait. The shaded leg depicts the swing phase ending at heel strike (HS) and the stance phase continuing until toe off (TO). The hip extends at the beginning of swing and continues through midstance phase.

Standing

The weight placed upon the hip joint in standing has been calculated. Using a lever system, as illustrated in Figure 9–11, the distance from the body weight center of gravity is approximately 4 inches to the center of fulcrum at the hip joint. The abductor gluteal muscles must exert a balancing force about the fulcrum through a lever arm of 2 inches. It is thus apparent that a 150-pound individual imposes a pressure of 450 pounds upon the femoral head. This force, which is normally sustained by the femoral head, explains why such tremendous forces are required to cause fracture and why the engineering studies of artificial hips have been so difficult. In diseased hips, added body weight of the individual must also be carefully controlled, as each pound of weight gain imposes three pounds upon the hip. A weight gain of 50 pounds would impose 150 pounds upon the hip.

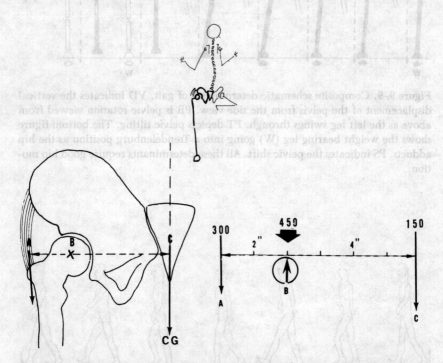

Figure 9–11. Weight borne by the hip on standing. *Left,* Weight upon the hip joint (B) combined from body weight (C) and balancing pull of the abductor muscles (A) (CG—center of gravity). *Right,* A 150-lb adult with a 4-inch distance from center of gravity to hip joint fulcrum is balanced by gluteal muscular action 2 inches from the fulcrum. The glutei exert 300 lb to balance; thus, total pressure upon the hip joint is 450 lb.

Use of a crutch or cane in the opposite hand markedly decreases the pressure upon the hip, as described in Figure 9–12. Assuming that the patient holds the cane 20 inches from the center of gravity and presses with 30 pounds of pressure, the pressure upon the hip joint (of a 150-pound person) is only 120 pounds, because there is less need for gluteal muscular balance. The cane acts in a clockwise direction, as do the gluteal muscles; thus, the cane and gluteal muscles balance each other.

PAIN IN THE HIP JOINT

There are four specific structures about the hip joint than can elicit pain: the fibrous capsule and its ligaments, the surrounding muscles, the bony periosteum, and the synovial lining of the joint. The cartilage is insensitive, but there is question as to the insensitivity of the subchondral bone because of its blood supply and its sensory vasomotor nerves.

The hip joint has sensory innervation from the femoral, obturator, superior gluteal, and accessory obturator nerves.

The branches of the femoral nerve supply the iliofemoral ligament near its femoral attachment. The articular branch of the obturator supplies the medial portion of the capsule. The superior gluteal nerve gives branches to the fibrous layer of the superolateral region of the capsule. When the accessory obturator nerve is present, it supplies the area supplied by a branch from the femoral nerve. The sciatic nerve is considered to supply the posterior aspect of the hip joint capsule via muscular branches to the quadrate and gemellus muscles.

The nerves that supply the hip joint also supply the muscles about the hip joint.

1. Femoral nerve—to the quadriceps muscle
2. Obturator nerve—to the external obturator muscle and the adductor muscle group
3. Inferior gluteal nerve—to the gluteus maximus muscle
4. Sciatic nerve—semimembranous and semitendinous muscles as well as the great adductor muscle, the gemellus muscle, and the quadrate muscle of thigh
5. Sacral plexus (S_1 and S_2)—to the piriform muscle, internal obturator muscle, gemellus muscles, and quadrate muscle of thigh
6. Superior gluteal nerve—to the gluteus medius and minimus muscles
7. Lumbar plexus (L_2 to L_4)—to the greater psoas muscle

The nerves supplying the short muscles of the hip (i.e., the obturator nerve, the sciatic nerve, and the sacral plexus) also supply the sensory

Use of a crutch or cane in the opposite hand markedly decreases the pressure upon the hip, as described in Figure 9–12. Assuming that the patient holds the cane 20 inches from the center of gravity and presses with 30 pounds of pressure, the pressure upon the hip joint (of a 150 pound person) is only 120 pounds, because there is less need for gluteal muscular balance. The cane acts in a clockwise direction, as do the gluteal muscles. The cane and gluteal muscles balance each other.

PAIN IN THE HIP JOINT

There are no specific structures about the hip joint than can elicit pain: the fibrous capsule and its ligaments, the surrounding muscles, the periosteum, and the synovial lining of the joint. The cartilage is insensitive, but the question as to the sensitivity of the subchondral trabeculae remains a good supply and its sensory vasomotor nerves.

The hip joint sensory innervation from the femoral, obturator, superior gluteal and accessory obturator nerves.

The branches of the femoral nerve supply the iliofemoral ligament near its femoral attachment. The medial branch of the obturator supplies the medial portion of the capsule. The superior gluteal nerve gives fibers to the fibrous layer of the superolateral region of the capsule. The accessory obturator nerve, if present, it supplies the area supplied by a branch from the femoral nerve. The sciatic nerve is considered to supply the posterior aspect of the hip joint capsule via muscular branches to the quadrate and gemellus muscles.

The nerves that supply the hip joint also supply the muscles that move the joint.

Femoral nerve—to the quadriceps muscle.
Obturator nerve—to the external obturator muscle and the adductor group.
Inferior gluteal nerve—to the gluteus maximus muscle.

as well as the great adductor muscle, the gemellus muscle, and the quadrate muscle of thigh.

Sacral plexus (S₄ and S₅)—to the piriform muscle, internal obturator muscle, gemellus muscles, and quadrate muscle of thigh.
Superior gluteal nerve—to the gluteus medius and minimus muscles.
Lumbar plexus (L₂ to L₄)—to the greater psoas muscle.

The nerves supplying the short muscles of the hip (i.e., the obturator nerve, the sciatic nerve, and the sacral plexus) also supply the sensory

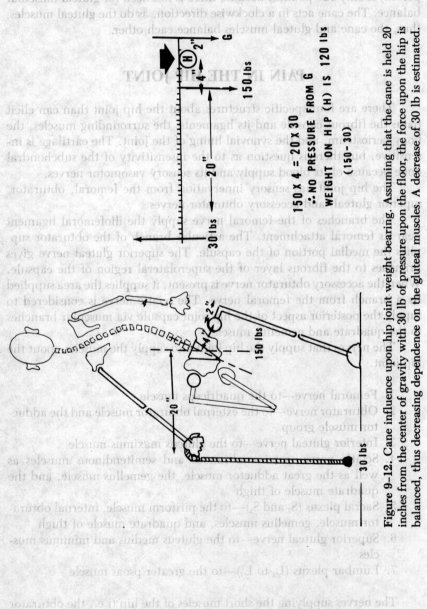

Figure 9–12. Cane influence upon hip joint weight bearing. Assuming that the cane is held 20 inches from the center of gravity with 30 lb of pressure upon the floor, the force upon the hip is balanced, thus decreasing dependence on the gluteal muscles. A decrease of 30 lb is estimated.

branches from the capsule. The cutaneous branches about the hip originate at a higher level than the motor and capsular nerves. The lateral
femoral cutaneous nerve covers the anterolateral thigh by L_2, and the anterior thigh by the continuation of the femoral nerve, L_2 to L_4. The upper
portion of the thigh is supplied by the iliohypogastric nerve, and the buttocks are supplied by the posterior primary division of D_{12} (T_{12}) to L_3. This
pattern of innervation implies that superficial cutaneous abnormality is
referred from higher spinal levels. It also explains why capsular irritation
via the obturator and sciatic nerves frequently refers pain distally into the
knee.

Pain in the hip region must be differentiated based on the specific
tissues involved. Because the hip is a weight-bearing joint and instrumental in ambulation, the sequence of pain is such that pain initially occurs
during standing or walking (weight bearing), and progresses to pain on hip
motion without weight bearing, to pain at rest. Pain is gradually followed
by limited range of motion in the hip. Because the cartilage of the hip is
avascular and lacks sensory fibers, pain must originate from other tissues.

Degenerative Arthritis

The most common painful condition of the hip joint is degenerative
joint disease, termed **osteoarthritis.** The exact cause and mechanism of
degenerative arthritis remain elusive, but much is being learned about the
disabling joint condition.

Arthrosis rather than arthritis might be a better term for this condition, as inflammation, implied by "itis," is not necessarily found. The
changes found are reparative and thus conform to the meaning of the suffix "osis." Deterioration of the articular cartilage and gradual subchondral
sclerosis occurs, with bone remodeling and osteophyte formation.

Trauma is considered the major cause of this osteoarthrosis. Aging is
also implicated, but this concept remains vague. Systemic factors have
been implicated, as has congenital predisposition. Joint cartilage degeneration has been claimed to be enzymic with superimposed trauma.

In degenerative arthritis, changes are noted in the subchondral bone.
Such changes include microscopic linear fractures, which may be the fore-
runner of ultimate cartilage degeneration.

Mechanical stress considered to be the most damaging[1] is repetitive
impulsive loading with sheer stresses that are not significant in a well-
lubricated joint. This consideration may indicate the basis for avoiding
this form of trauma as imposed by jogging and running on hard surfaces
for long distances and durations, even though a recent survey of sports
medicine has claimed that after years of running, long-distance runners
fail to show evidence of degenerative arthritis. It is possible that asympto-

matic long-distance runners have normal joints to begin with, which allow them to run for long periods of time, and that only when there is a mechanical defect in the hip does any running ultimately cause degenerative changes.

Mechanical trauma is now well known to be influenced by incongruity of the joint surfaces. Hip dysplasia in childhood, slipped femoral capital epiphysis, coxa plana, and other conditions that alter the plane and surface relationship in adult life increase susceptibility to mechanical trauma.

The articular cartilage is composed of four layers. The superficial layer consists of flattened chondrocytes surrounded by tightly woven bundles of collagen fibers lying parallel to the subchondral bone. The intermediate zone has random arrangement of collagen fibers intertwining around chondrocytes. In the deeper midlayer, the collagen fibers are perpendicular to the surface. This layer permits compression and resiliency of the tissue. The deepest layer is the transition layer between cartilage and bone. It is calcified with minimal fibers and cells.

The cartilage is avascular and depends on imbibition and diffusion for its nutrition. Much of the nutrition is afforded by the synovial fluid. Diffusion of nutritive fluids through the endochondral plates via the blood vessels is accepted but not yet fully understood.

Lubrication of cartilage upon cartilage is considered to occur from a glycoprotein faction that is pumped out of cartilage by pressure. Motion and intermittent compression of cartilage is thus mandatory for adequate nutrition. Cartilage degenerates because of subchondral sclerosis, which impairs diffusion. Thus, a decrease in the lubricant occurs, which enhances superficial damage.

It has been postulated that repeated forceful perpendicular impact on the limb causes microfractures of the cancellous endochondral bone, which, as they heal, occlude the permeability of nutritive fluids to the cartilage and thus predispose the joint to osteoarthritis.

The cartilage itself does not have a nerve supply. The sensory nerve endings are found in the capsule and the ligaments, but none penetrate the intact cartilage. The synovial membrane is well innervated, but most nerves innervate the blood vessels, and very few are sensory for pain transmission.

The synovial membrane is probably involved early in arthralgia with hyperemia. Ultimately, the membrane increases in both quantity and composition of synovial fluid. This synovial inflammation affects the autonomic nerves, which include sensory and vasomotor nerves, so that pain and further congestion result. Synovium has been shown to be sensitive to hyperemia and distension.

The capsule also plays a vital part in the production of pain. The capsule becomes infiltrated by inflammatory cells, and gradual thickening

Usually, the capsule is tense, but it is relaxed in flexion, abduction, and external rotation. The iliofemoral and the ischiofemoral ligaments become taut in hip extension, adduction, and internal rotation. Since innervation of external rotator (the gemellus and quadrate muscles) is closely related to the innervation of the capsule, it becomes of clinical significance that in hip pathology involving the capsule, flexion remains unrestricted for a longer period than do rotation and hip extension. As the joint surfaces change in contour and depth, abduction and external rotation become increasingly more difficult, with progressive limitation and proportional pain.

Circulatory impairment is also considered to cause hip pain. These vascular changes occur not only in the articular and periarticular tissues but also in the bone and subchondral tissues. The complete relationship of vascular impairment to hip degeneration and pain is not yet clear.

Management of Hip Pain

The usual effective measures to decrease hip pain can be classified as nonsurgical and surgical.

Nonsurgical Measures. The nonsurgical measures are rest, immobilization by casting, traction, intra-articular injection of steroids or an anesthetic agent, and chemical denervation.

Rest helps minimize or eliminates weight bearing. Prescribed rest can vary, ranging from restricted standing or walking to elimination of jumping or running, or to complete bedrest. Use of a cane in the opposite hand or crutch(es) also decreases weight bearing. Loss of excessive body weight is always indicated.

Proper positioning during prolonged bedrest must be assured. Full range of motion must be maintained during bedrest to prevent contracture weakness or atrophy. Flexion contracture is by far the most common problem; therefore, it should be prevented.

Lying in the prone position frequently during the day helps stretch the hip flexors. A pillow placed under the thigh with the patient in the prone position is of value, providing it does not cause excessive lumbar lordosis. Hip extensor exercise to strengthen the extensor musculature is also valuable (Figs. 9–13, 9–14, and 9–15).

Immobilization by casting rests the inflamed joint. A spica cast usually is effective and must include the pelvis. Keeping the knee free and exercising the quadriceps muscle during cast application are indicated.

Traction is effective in immobilizing the hip joint and elongating the capsule of the hip, thus separating the articular surfaces. This traction can be supplied by pin insertion or by skin application (Fig. 9–16).

Intra-articular injection of a steroid or anesthetic agent requires special skill. Usually, the joint can be entered by measuring 2 to 3 cm below

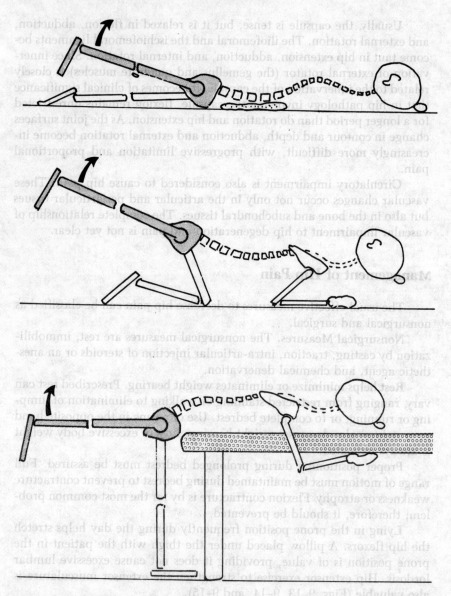

Figure 9–13. Exercises to extend hip joint: exercises aimed to stretch the anterior hip capsule and strengthen the extensor musculature. *Top,* With patient prone, the leg is extended, preferably against resistance. A pillow under the abdomen decreases excessive lordosis. *Center,* With contralateral knee flexed and bearing weight, the involved leg is extended. *Bottom,* Exercise is similar to that in center illustration except that the patient is prone over table. The dependent leg stabilizes the pelvis, and the lumbar lordosis is decreased.

Figure 9–14. Hip flexor stretching exercise. With patient in supine position and the normal hip held to the chest, the opposite leg by its own weight or weighted by sandbag is extended actively and passively.

the anterosuperior iliac spine and entering 2 to 3 cm lateral to the femoral artery (as identified by pulsation) (Fig. 9–17). The needle proceeds posteriorly and medially at a 60-degree angle until the capsule is penetrated (Fig. 9–18). Aspiration should precede injection. Fluoroscopy may be helpful.

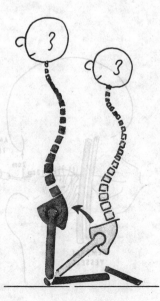

Figure 9–15. Hip extensor exercise. From the full kneeling position, the patient arises to full erect kneeling posture. This stretches the hip flexors and strengthens extensors.

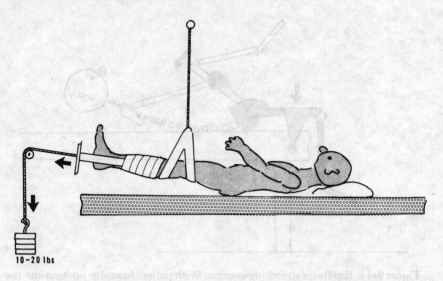

Figure 9–16. Technique of hip traction by skin application to lower leg.

the anterosuperior iliac spine and entering 2 to 3 cm lateral to the femoral artery (as identified by pulsation) (Fig. 9–17). The needle proceeds posteriorly and medially at a 60-degree angle until the capsule is penetrated (Fig. 9–18). Aspiration should precede injection. Fluoroscopy may be helpful.

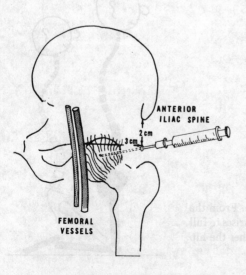

Figure 9–17. Technique and site of intra-articular injection. The needle is inserted 2 to 3 cm below the anterosuperior iliac spine and 2 to 3 cm lateral to the femoral artery pulsation and penetrates in a posterior medial direction (a 60-degree angle) until bone is reached. After aspiration, which may involve fluid, the steroid-anesthetic agent is injected.

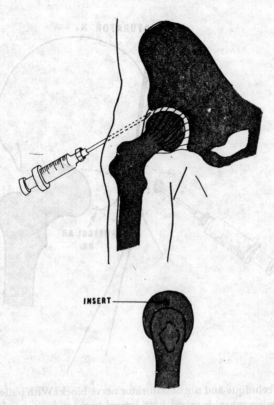

INSERT

Figure 9–18. Intra-articular injection. While patient is lying on the contralateral hip, the trochanter is palpated, and the needle is inserted midline of the superior aspect of the femur neck—to follow along neck until the capsule of the hip joint is reached. Aspiration of fluid is followed by a steroid-anesthetic agent.

Chemical denervation of the capsule requires nerve blocks using an anesthetic agent and most frequently involves the obturator nerve (Fig. 9–19).

Surgical Intervention. When conservative measures fail, surgical intervention may be considered. Procedures include the following:

1. Revascularization of bone procedures
2. Osteotomy to alter alignment of head to acetabulum
3. Denervation
4. Arthrodesis
5. Arthroplasty of cup or total hip replacement

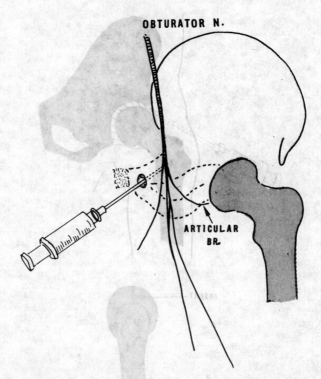

Figure 9–19. Technique and site of obturator nerve block. With patient supine and thighs separated, a wheal is raised 1 cm lateral to the pubic tubercle. A 22-gauge 8-cm needle is directed perpendicularly until the inferior ramus of the pubis is reached. The direction of the needle is then changed several centimeters into a lateral and superior direction, parallel with the pubic ramus. After aspiration, 5 to 10 ml anesthetic agent is injected. An effective injection is determined by adduction and external rotation paresis and *not* by an area of anesthesia.

PAIN FROM OTHER SITES

Pain claimed to be in the hip area by the patient may be referred there from other sites. This consideration must always be included in the differential diagnosis.

Myofascial pain from trigger areas within muscle, myofascial tissues, the fascia lata, or ligaments may occur (Fig. 9–20). These trigger areas may be local or distant, and they respond well to local injection of an anesthetic agent, deep massage, or even vasocoolant spray.

Pain may be referred to the hip (buttock) area from the lumbar spine. In this condition, the pain can be reproduced by lumbar movement. Usu-

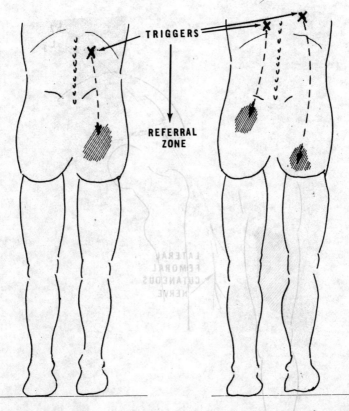

Figure 9–20. Pain referred to hip region. Radiating pain may occur from irritation of the nerves emanating from the thoracic spine, especially T_6 and T_4. Pressure that is paraspinous in relation to these levels elicits buttock pain.

ally, pain is reproduced more by hyperextension of the lumbar spine than by flexion. Pain and tenderness are usually elicited by pressure upon the sciatic nerve in the vicinity of the sciatic notch. Straight leg raising may be painful, and some sensory or motor deficit may be noted.

Entrapment of the lateral femoral cutaneous nerve, termed **meralgia paresthetica,** causes a burning pain in the anterolateral portion of the thigh (Fig. 9–21). Usually, the area is very clearly delineated, and ultimately the pain may be accompanied by numbness of the area. The lateral femoral cutaneous nerve (L_2 to L_3) lies within the pelvis and emerges superficially in the region of the anterosuperior iliac spine to lie beneath the deep lateral fascia of the thigh. Entrapment usually is at the lateral end of the inguinal ligament below the anterosuperior spine. Often, the history of trauma cannot be established.

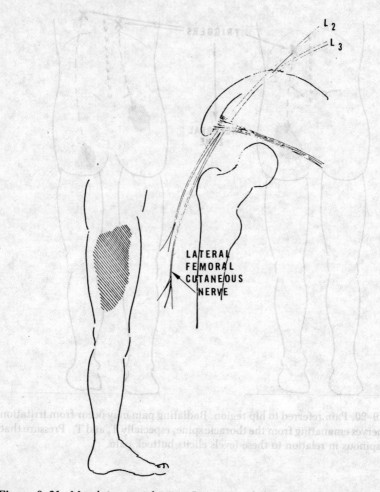

Figure 9–21. Meralgia paresthetica. Irritation of the lateral femoral cutaneous nerve at any point where along its course that is distal from L_2 to L_3 can cause a burning pain in the cutaneous zone depicted.

Postural exercises to decrease lordosis are considered to be valuable in treating meralgia paresthetica. Often, applying a heel lift on the opposite shoe relieves tension upon the fascia lata and affords relief. Oral anti-inflammatory medicine should be taken. Surgical decompression of the nerve is rarely indicated.

REFERENCE

1. Ravin, EL: Mechanical aspects of osteoarthrosis. Bull Rheum Dis 26(7):862–865, 1975–1976 series.

BIBLIOGRAPHY

Blount, WP: Don't throw away the cane. J Bone Joint Surg [Am] 38:695, 1956.

Cailliet, R: Foot and Ankle Pain, ed 2. FA Davis, Philadelphia, 1983.

Denham, RA: Hip mechanics. J Bone Joint Surg [Br] 41:550–557, 1959.

Gardner, E: The innervation of the hip joint. Anat Rec 101:353–371, 1948.

Inman, VT: Functional aspects of the abductor muscles of the hip. J Bone Joint Surg 29:607, 1947.

Lloyd-Roberts, GC: The role of capsular changes in osteoarthritis of the hip. J Bone Joint Surg [Br] 37:8–47, 1955.

Maistrelli, G, Gerundini, M, and Bombelli, R: The inclination of the weight bearing surface in the hip joint: A review paper. Orthopedic Review 15(5):23–31, 1986.

Ravin, EL: Mechanical aspects of osteoarthrosis. Bull Rheum Dis 26(7):862–865, 1975–1976 series.

Saunders, JB, Inman, VT, and Eberhart, HD: The major determinants of normal and pathological gait. J Bone Joint Surg [Am] 35:543–558, 1953.

Trueta, J, and Harrison, MHM: Normal vascular anatomy of femoral head in adult man. J Bone Joint Surg [Br] 35:442–461, 1953.

Wadsworth, JB, Smidt, GL, and Johnston, RC: Gait characteristics of subjects with hip disease. Phys Ther 52:829–837, 1972.

CHAPTER 10
Knee Pain

Of all the body joints, the knee is probably most vulnerable to becoming a source of pain. This vulnerability is due to the joint's role in gait and stance, and its greater importance in bending, stopping, and squatting. Because of its structure, the knee is unstable. It depends totally on ligamentous support and strong muscular function. It has an extensive synovial membrane. All these factors have been fully discussed in *Knee Pain and Disability*,[1] but are beneficial to review.

STRUCTURAL ANATOMY

There are two joints in the knee: the femorotibial and the femoropatellar (Fig. 10–1).

The distal end of the femur has two convex condyles separated inferiorly by a deep V-shaped notch and anteriorly by a concave depression into which fits the patella.

The femoral condyles articulate with the concave surface of the tibial plateau. This articulation forms the femorotibial joint. The articular surfaces are not symmetric, so they do not form a stable congruent joint. Symmetry is created by the interposition of fibrocartilaginous menisci that assist in distributing pressure between the femur and the tibia, increase the elasticity of the joint, and assist in its lubrication.

Ligaments

There are strong ligaments on the medial and lateral aspects of the joint extending from the femoral condyles to the tibia and fibula (Fig. 10–

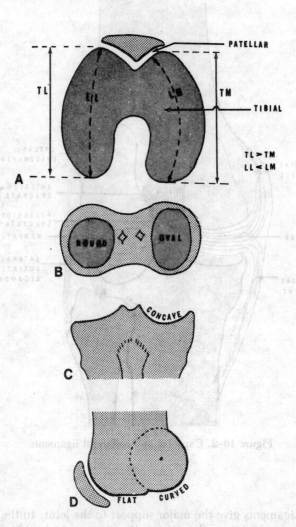

Figure 10-1. Knee joint surfaces. *A*, Femoral condyle surfaces of the right knee. TL—anteroposterior length of the lateral condyle; TM—length of the medial condyle. The medial condyle (LM) is longer than the lateral condyle (LL) because of its curved surface. *B*, Superior surface of the right tibia. The lateral articular surface is rounded, and the medial articular surface is oval. *C*, The medial tibial articular surface is deeper and more concave than the lateral. *D*, Side view of the femur showing the flat anterior surface and the curved posterior surface. The two articulations are illustrated in *A*: the patellar, in which the patella articulates with the femur, and the tibial, which then glides upon the tibia.

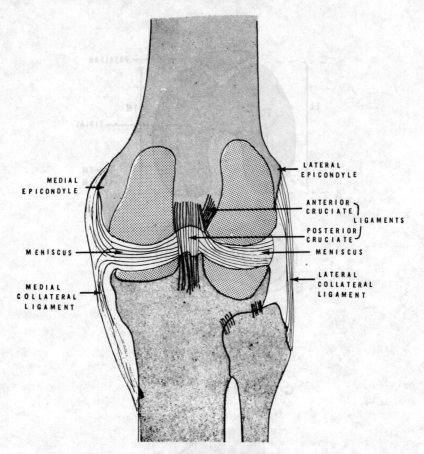

Figure 10–2. Capsular and collateral ligaments.

2); these ligaments give the major support to the joint. In the centrum of the tibiofemoral joint are the cruciate ligaments, which add to the stability and assist in the normal mechanical function of the knee.

The fibrous capsule of the joint has selective thickenings that form ligaments.

Medial Capsular Ligaments. The medial capsular ligaments divide into deep and superficial portions (Fig. 10–3). The deep portion is further divided into three sections: anterior, middle, and posterior. The anterior fibers extend anteriorly into the extensor mechanism. The middle fibers stabilize the joint against lateromedial motion and penetrate the joint to attach to the medial meniscus. The posterior fibers extend into the formation of the posterior popliteal capsule.

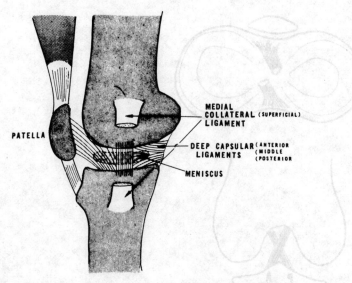

Figure 10-3. Medial capsular ligaments. The superficial collateral ligament attaches superiorly to the medial femoral condyle and attaches onto the tibia below the articular cartilage. The deep capsular ligament divides into three portions: anterior, middle, and posterior.

The superficial medial ligament is more distinct and forms the medial collateral ligament. It is attached superiorly to the medial femoral epicondyle and inferiorly upon the tibia just below the level of the articular cartilage. There are numerous bursae between the deep and the superficial capsular ligaments.

Lateral Capsular Ligaments. The fibular collateral ligament, a distinct thickening of the capsule, passes from the lateral epicondyle of the femur to the head of the fibula. In its course, it is surrounded by the dividend tendons of the biceps and is penetrated by the popliteus tendon as the tendon passes to attach to the lateral epicondyle of the femur. The peroneal nerve passes the neck of the fibula behind the biceps tendon in the region.

The knee has its maximum stability at full extension when the collateral ligaments are taut. Immediately upon flexion, the collateral ligaments relax and permit lateromedial motion and rotation of the tibia upon the femoral condyles.

Cruciate Ligaments. The paired cruciate ligaments are named according to their tibial attachment (Fig. 10-4). The anterior ligament origi-

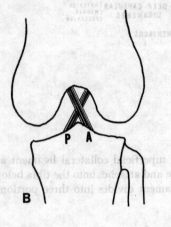

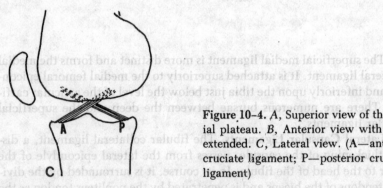

Figure 10-4. A, Superior view of the tibial plateau. B, Anterior view with knee extended. C, Lateral view. (A—anterior cruciate ligament; P—posterior cruciate ligament)

nates from the anterior tibial plateau and proceeds superiorly and posteriorly to attach to the medial aspect of the lateral femoral condyle. The posterior ligament arises from the posterior aspect of the tibia and extends forward, upward, and inward to attach to the medial femoral condyle.

By their attachments and direction, they restrict shear motion of the joint and thus act in flexion and extension of the knee. The anterior cruciate ligament prevents knee hyperextension, and the posterior cruciate ligament mechanically assists the knee in flexion (Fig. 10-5).

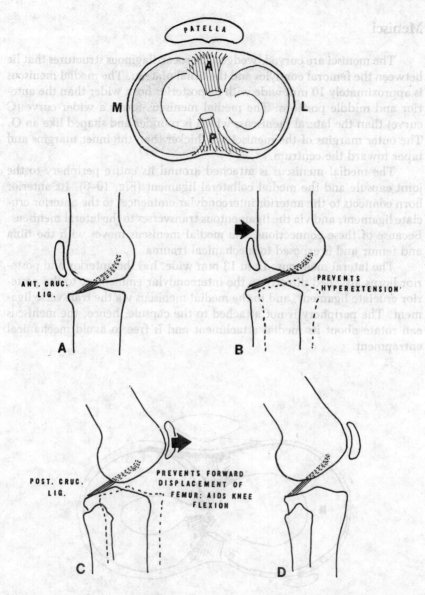

Figure 10–5. Cruciate ligament function. *Top*, Superior view showing the cruciate origin and direction. *A* and *B*, Lateral view showing anterior cruciate as it prevents hyperextension. *C* and *D*, Posterior cruciate ligament function, which prevents lateral displacement of the tibia upon the femur and aids in normal knee flexion.

Menisci

The menisci are curved, wedged, fibrocartilaginous structures that lie between the femoral condyles and the tibial plateau. The medial meniscus is approximately 10 mm wide with its posterior horn wider than the anterior and middle portions. The medial meniscus forms a wider curve (C curve) than the lateral meniscus, which is rounder and shaped like an O. The outer margins of the menisci are thicker than the inner margins and taper toward the centrum.

The medial meniscus is attached around its entire periphery to the joint capsule and the medial collateral ligament (Fig. 10-6). Its anterior horn connects to the anterior intercondylar eminence, to the anterior cruciate ligament, and via the ligamentous transverses to the lateral meniscus. Because of these connections, the medial meniscus moves with the tibia and femur and is exposed to mechanical trauma.

The lateral meniscus, 12 to 13 mm wide, has the anterior and posterior horns attached directly to the intercondylar eminences, to the posterior cruciate ligament, and to the medial meniscus via the transverse ligament. The periphery is not attached to the capsule; hence, the meniscus can rotate about its medial attachment and is free to avoid mechanical entrapment.

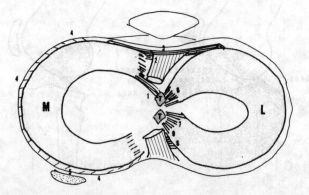

Figure 10-6. Attachment of the menisci. Right tibial plateau viewed from above. 1, Fibrous attachment of medial meniscus (M) to tibial tubercle (T). 2, Connection to anterior cruciate. 3, Transverse ligament, which connects to the anterior horn of the lateral meniscus (L). 4, Meniscus (M) is attached around the entire periphery to the capsule. 5, Attachment to semimembranous muscle tendon. 6, Lateral meniscus anterior horn, and 7, posterior horn, attached to eminentia intercondylaris (T) and attached to posterior cruciate ligament (8). 9, A fibrous band attaches superiorly into the fossa intercondylaris at the femur.

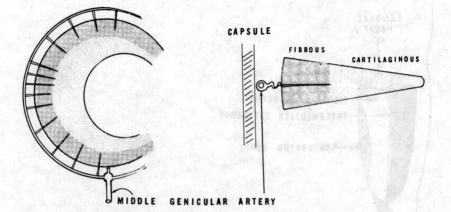

Figure 10-7. Intrinsic circulation of the menisci. The middle genicular branch of the popliteal artery sends branches around the periphery of the meniscus under the capsule. Small tortuous nonanastomotic vessels enter the outer fibrous zone of the meniscus. Their tortuosity permits movement of the meniscus. The inner third of the meniscus is cartilaginous and avascular.

Statistics have substantiated that the medial meniscus sustains injury more often than the lateral meniscus. Studies cite ratios ranging from 3:1 to 20:1. This finding is partially attributed to the attachment variation.

The menisci have a unique intrinsic blood supply (Fig. 10-7). The middle genicular artery, a branch of the popliteal artery, branches circuitously around the menisci and sends small tortuous vessels into the outer third of the menisci. The middle and inner thirds of the menisci are avascular. This vascular factor accounts for repair of meniscus injuries to the outer third (fibrous portion) and the failure to heal in the inner two thirds (avascular cartilaginous portion).

Nerves

The knee joint has a rich sensory innervation that can transmit pain. All of the following tissues are innervated by the same nerves: the skin, the synovial membrane, the capsule, the ligaments, the muscles, and the bursae.

The skin is supplied primarily by the femoral and the obturator nerves (Fig. 10-8). There is a minor supply by the sciatic nerve. The synovial capsule is a relatively insensitive tissue, and the articular cartilage carries no sensory fibers.

The fibrous capsule and the ligaments are richly supplied by medullated and nonmedullated afferent somatic nerves capable of carrying pain

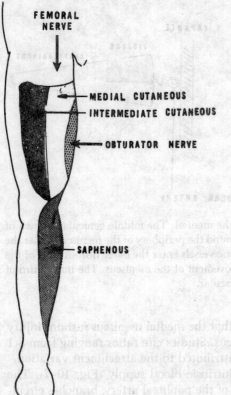

FEMORAL
NERVE

←── MEDIAL CUTANEOUS
←── INTERMEDIATE CUTANEOUS

←── OBTURATOR NERVE

←── SAPHENOUS

Figure 10-8. Cutaneous regions of femoral nerve. The dermatomal distribution of the femoral nerve is depicted in the shaded areas.

sensation. Some of these articular nerve fibers penetrate the synovial membrane and can elicit pain from these regions. The capsular and ligamentous structures are innervated by the sciatic nerve (articular branch to the lateroposterior area). The tibial articular branch supplies the posterior aspect of the joint, and the external popliteal nerve supplies the lateral articular area. The obturator nerve also sends a small branch to the posterior capsule. The anteromedial aspect of the capsule is supplied by the femoral nerve (Fig. 10-9).

The arterioles of the synovium are supplied by autonomic fibers and have sensory somatic fibers. Thus, vascular changes initiate somatic changes; this can explain the marked painful reactions to heat, cold, or barometric pressure changes claimed by arthritic patients.

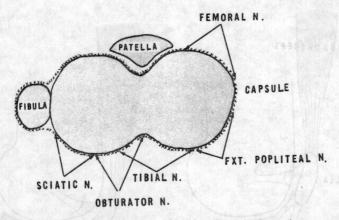

Figure 10-9. Cutaneous nerve distribution of knee area. Viewed from above, the knee capsule receives skin innervation from the sciatic, obturator, tibial, femoral, and external popliteal nerves. Pain can be elicited from capsular irritation at any of these areas.

Muscles

The knee joint is stabilized and powerfully motored by muscles that cross the joint from their origin above the hip joint and from the shaft of the femur to insert upon bony structures below the knee joint. These muscle groups are commonly classified as extensors (anterior), flexors (posterior), adductors (medial), and abductors (lateral). The abductors and adductors act upon the knee as rotators only when the joint is flexed.

The extensors that are of greatest importance in the stability and function of the knee are the quadriceps femoris muscles. They are composed of four heads: the rectus femoris, vastus medialis, vastus lateralis, and vastus intermedius. The rectus femoris muscle originates from the anterior iliac spine and all the vastus muscles from the shaft of the femur. All four heads converge into a common tendon that crosses the knee joint to attach to the tibial tubercle.

Within the patellar tendon lies the patella, which provides mechanical leverage to the extensor mechanism and provides a gliding surface against the femur to minimize friction. The quadriceps muscle is innervated by the femoral nerve formed by the anterior division of L_2, L_3, and L_4.

Besides extending the tibia upon the femur, the extensor mechanism also has ligamentous attachments to the menisci (Fig. 10-10) that permit the menisci to move during knee motion to prevent entrapment.

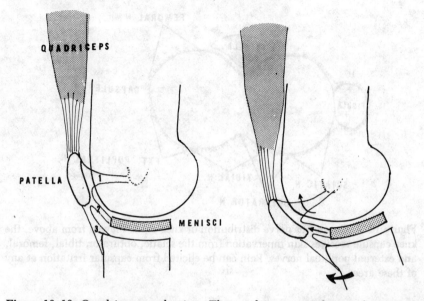

Figure 10-10. Quadriceps mechanism. The quadriceps extends over the anterior knee joint with three ligamentous extensions: 1—the epicondylopatellar portion, which attaches to the epicondyle eminence of the femur and guides rotation of the patella; 2—the meniscopatellar portion, which attaches to and pulls the meniscus forward during knee extension; and 3—the infrapatellar tendon, which attaches to the tibial tubercle and extends the tibia upon the femur.

The flexors are in the posterior aspect of the femur (Fig. 10-11). They allow the knee to cross and the leg to flex upon the thigh and rotate. The flexors are best divided into medial and lateral groups, with the medial containing the semimembranous muscle, the semitendinous muscle, and the lateral biceps muscle of the thigh. With the knee flexed, the medial group rotates the leg internally and the biceps rotates the leg externally.

All the flexors originate from a common site on the ischial tuberosity. The semitendinous muscle descends the medial aspect of the thigh, and as it crosses the knee, it joins the sartorius and gracilis muscles to form a common tendon, the pes anserinus (Fig. 10-12). This tendon flexes the knee and internally rotates the flexed leg. A bursa separates this tendon from the underlying femoral condyle and is clinically significant in that it can cause pain.

The semimembranous insertion divides into four tendons that blend into the capsule, but it has a deep fibrous extensor that attaches to the medial meniscus to pull it posteriorly as the knee flexes (Fig. 10-13).

The biceps femoris tendon attaches to the head of the fibula by three fibrous insertions. One insertion flexes the knee and externally rotates the

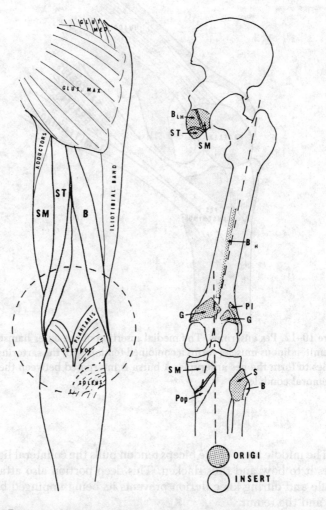

Figure 10-11. Posterior thigh muscles: flexors. *Left,* Semimembranous (SM), semitendinous (ST), and biceps muscle of thigh (B). The other muscles are labeled. *Right,* The origin and insertion of the posterior muscle groups.

B$_{LH}$—biceps long head
B$_{SH}$—biceps short head
B—biceps
S—sartorius
Pl—plantar
Pop—popliteal
G—Heads of gastrocnemius

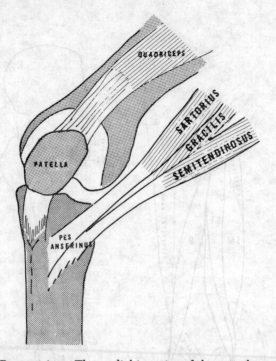

Figure 10-12. Pes anserinus. The medial insertion of the outer hamstring muscle, the semitendinous muscle, forms a conjoined tendon with the sartorius and gracilis muscles to form the pes anserinus. A bursa is interposed between the tendon and the femoral condyle.

leg. The middle layer of the biceps tendon pulls the collateral ligament and causes it to bow and thus slacken. This deep portion also attaches to the capsule and during knee flexion prevents its being impinged between the tibia and the femur.

The flexor muscle groups are innervated by the sciatic nerve. As the sciatic nerve divides into the tibial nerve and the common peroneal nerve, the former innervates the semimembranous muscle, semitendinous muscle, long head of the biceps muscles, and the common peroneal nerve innervates the short head of the biceps muscle.

The synovial capsule of the knee joint is large (Fig. 10-14), holding as much as 40 ml of air before it distends. Anteriorly, it ascends to two finger-breadths above the patella; posteriorly, it ascends to the origin of the gastrocnemius muscle; laterally, it extends superiorly to the margin of the epicondyles and inferiorly to 1 cm below the articular margin where the collateral ligaments attach.

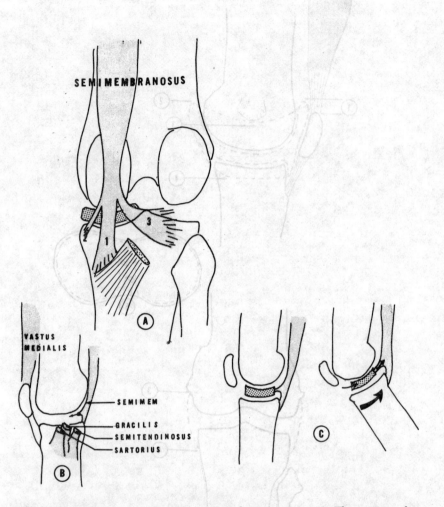

Figure 10-13. Medial aspect of the posterior knee structure. *A*, The semimembranous muscle has four tendinous inserts. The major insert, (1) extends to attach on the posterior aspect of the tibia and sends fibers into the popliteus muscle. In its path is an exterior branch that attaches to the posterior aspect of the medial meniscus (2 and 3). These tendons complete the posterior popliteal fossa and tense the capsule. *B*, The medial aspect of the knee with the insertion sites of the medial flexors. *C*, The semimembranous muscle flexes the knee and simultaneously pulls the meniscus backward and rotates it with the tibia.

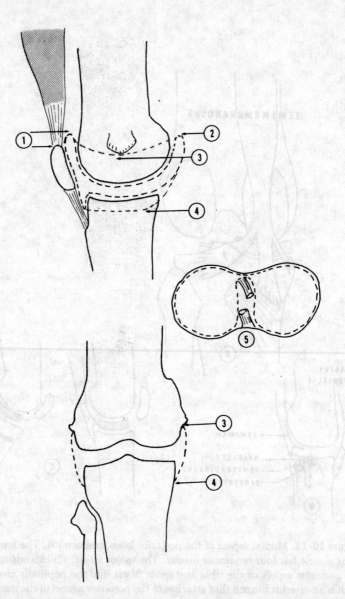

Figure 10-14. Synovial capsule. The capsule anteriorly ascends to two finger-breadths above the patella (1); posteriorly it ascends to the origin of the gastrocnemius muscle (2); laterally it attaches to the femur at the epicondylar level (3); inferiorly it attaches upon the tibia ¹/ı inch below the articular margin just below attachment of the collateral ligament (4). The cruciate ligaments (5) invaginate the capsule but are extracapsular.

FUNCTIONAL ANATOMY

The knee flexes and extends in an intricate manner that predisposes it to damaging stresses when violated. Knee flexion and extension are gliding movements of the tibia upon the femur with simultaneous rotation. During knee flexion, the lower leg rotates internally upon the femur, and during knee extension, it rotates externally.

Because of the contour of the femoral head with its anterior surface flat and its posterior position curved, the first 20 degrees of flexion involves a rocking motion followed by gliding until the tibial surface rotates about the posterior curved femoral condyles (Fig. 10–15). Further flexion is assisted by the posterior cruciate ligaments, which, once fully elongated, create a fulcrum about which the tibia moves.

Once flexion begins, the capsular ligaments relax to permit rotation. Most rotation occurs during the final phases of full flexion and full extension, but some degree of rotation occurs throughout flexion and extension. With the knee flexed to 90 degrees, the lower leg can rotate 40 degrees. With the knee fully extended, the joint surfaces are in direct opposition, and the collateral ligaments are taut, so that no rotation or lateral deviation of adduction or abduction is possible. The fully extended knee is stable and resistant to stress.

The direction and extent of rotation depends on the anatomic configuration of the articular surfaces. The medial femoral condyle is longer in its curved axis than is the lateral condyle, so as extension or flexion upon the lateral articular surface is reached, there is some remaining gliding surface upon the medial condyle, hence causing external rotation into further extension and internal rotation into further flexion (Fig. 10–16).

Muscular action also assists rotation at the knee. The quadriceps group is medially oblique across the knee joint and rotates the tibia externally during extension. The popliteal muscle begins internal rotation of the tibia during initiation of flexion.

The menisci are fixed to the tibia and femur and thus move during knee flexion and extension. In flexion-extension, the menisci move with the tibia whereas in rotation, the menisci move with the femur upon the tibia. The menisci are not subject to impingement if the flexion-extension motion pattern is not violated.

The posterior cruciate ligament assists in flexion. The cruciate ligaments also crisscross and thus limit rotation of the tibia upon the femur. The posterior cruciate ligament prevents excessive internal rotation, and the anterior cruciate ligament prevents excessive external rotation. Abnormal shearing forces or rotational forces upon the knee can cause disruption of the cruciate ligaments as well as damage to the menisci once the capsular ligaments are overstretched or torn.

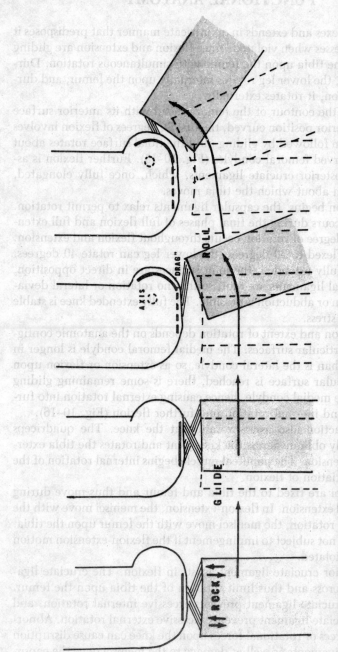

Figure 10–15. Mechanism of knee flexion. The first 20 degrees of flexion consists essentially of a rocking motion followed by the femur gliding upon the tibia. Rotation (flexion) occurs when the tibia reaches the rounded posterior femoral condyle. The posterior cruciate ligament acts as a drag to assist flexion.

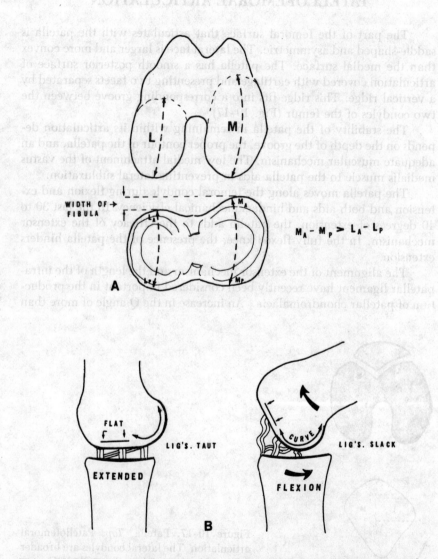

Figure 10–16. Passive mechanism of knee flexion-extension. *A,* The articular surface of the medial condyle is longer than that of the lateral (M_A to M_P greater than L_A to L_P); thus the tibia travels further upon the medial condyle during extension and rotates externally. *B,* Because of the flat anterior femoral surface, the collateral ligaments are taut and the knee is stable. The ligaments relax upon flexion.

PATELLOFEMORAL ARTICULATION

The part of the femoral surface that articulates with the patella is saddle-shaped and asymmetric. The lateral face is larger and more convex than the medial surface. The patella has a smooth posterior surface of articulation covered with cartilage and presenting two facets separated by a vertical ridge. This ridge fits into a corresponding groove between the two condyles of the femur (Fig. 10–17).

The stability of the patella in remaining within its articulation depends on the depth of the groove, the proper contour of the patella, and an adequate muscular mechanism. The low medial attachment of the vastus medialis muscle to the patella aids in preventing lateral subluxation.

The patella moves along the femoral condyle during flexion and extension and both aids and hinders mechanical efficiency. In the last 30 to 40 degrees of extension, the patella adds to the efficacy of the extensor mechanism. In the fully flexed knee, the presence of the patella hinders extension.

The alignment of the extensor mechanism and the length of the infrapatellar ligament have recently been considered important in the production of patellar chondromalacia. An increase in the Q angle of more than

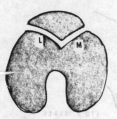

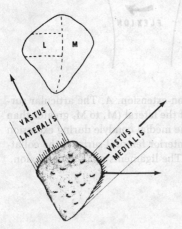

Figure 10–17. Patella. *Top*, Patellofemoral articulation. The lateral condyles are broader and more concave than the medial. *Center*, Facet planes of the patella on the articular surface: three planes on the lateral half and one plane on the medial. *Bottom*, Muscle pull upon the patella with the vastus lateralis muscle pulling cephalad. The vastus medialis muscle, by attaching lower and more laterally, exerts medial pull, thus centralizing and stabilizing the patella.

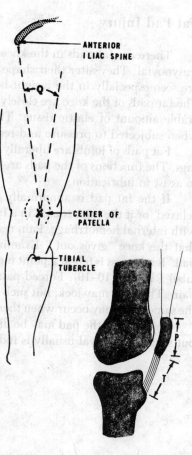

Figure 10-18. Q angle—patellotendon ratio. The Q angle is drawn from a line originating at the anterior iliac spine to the center of the patella (essentially the direction of pull of the quadriceps) and a line from the center of the tibial tubercle to the center of the patella. A normal Q angle is considered to be 20 degrees or less. Roentgenographic measurement of the length of the patella (P) compared to length of infrapatellar ligament (T) determines the height the patella rides. T should equal P. If T is longer, it indicates a high-riding patella.

20 degrees and an increased length of the patella have been considered conducive to chondromalacia (Fig. 10–18).

PAIN SYNDROMES

Complaints of pain in the knee must be clarified by clinical manifestations. Whereas the diagnoses of low back pain and cervical pain are based primarily on the history, knee pain requires a careful examination. Pain causes the patient to seek help and localizes the site, but a careful examination is mandatory.

Steindler[2] attempted further classification of pain localized anteriorly, anteromedially, and anterolaterally in the knee into the specific structures involved (Fig. 10–19). He regarded trigger points, or the sites of maximum tenderness, as being valuable in differentiation of these specific structures.

Fat Pad Injury

There are fat pads in the knee joint that are intracapsular, albeit extrasynovial. They alter their shape with movement of the joint. Fat pads are seen especially in the weight-bearing lower extremities of mammals. The fat pads of the knee are closely packed fat cells contained in a considerable amount of elastic tissue. This tissue causes a firm pad to deform when subjected to pressure and regain its shape upon release.

Fat pads of joints are liberally supplied with pain receptor nerve endings. The functions of the pads are considered to be to fill dead spaces and to assist in lubrication.

If the fat pad is abnormally large, if the quadriceps mechanism is relaxed, or if the joint is used in a faulty manner, the pad can be impinged with internal hemorrhage. Pain results, and frequently, the patient claims that the knee "gives out." Examination may reveal hypertrophy of the pad. Tenderness is felt at a point medial (lateral) to the patellar tendon (see label 1 in Fig. 10-19). Forced passive extension of the knee increases the pain. The knee may lock, but such locking is unusual. Bilateral fat pads in the same knee may occur when there is concomitant degenerative arthritis.

Injection of the pad may be diagnostic and momentarily therapeutic, but surgical removal usually is indicated.

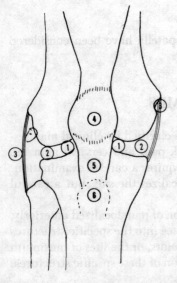

Figure 10–19. Tender sites of anterior aspect of the knee. 1—the site of painful fat pads; 2—meniscus sites of tenderness; 3—collateral ligament pain (medial and lateral); 4—patellar pain and tenderness (see text for explanation); 5—infrapatellar bursal pain; 6—tibial tubercle (Osgood-Schlatter disease).

Meniscus Injury

As previously stated, the medial meniscus sustains injury more often than the lateral meniscus.

The mechanisms causing meniscus injury are numerous, but are predominantly compression and traction. The usual injurious mechanism is rotatory stress on the weight-bearing leg. The stress is imposed by violation of internal rotation during flexion or external rotation during extension. These motion patterns are physiologic.

During knee flexion and extension, the menisci move anteriorly and posteriorly, respectively. With maximum flexion, the posterior portion of the meniscus is compressed between the posterior surfaces of the tibial and femoral condyles. Rotation of the tibia upon the femur in this flexed position displaces the posterior horn of the meniscus toward the center of the joint. Forceful extension of the knee either tears the posterior attachment of the meniscus or causes a longitudinal tear in the meniscus (Fig. 10–20). The direction of rotation determines the specific meniscus entrapped and the specific type of tear (Fig. 10–21).

Tearing of a meniscus with the knee fully extended is rare unless the stress is violent and disrupts the collateral or cruciate ligaments with or without a condylar fracture. Meniscus injury requires flexion and extension of the knee *combined* with inappropriate rotation when the lower leg is fixed to the ground in a weight-bearing position.

Another mechanism considered to cause injury to a meniscus is externally forced valgus of the knee during flexion and external rotation that excessively although momentarily opens the joint space and consequently entraps the meniscus.

A complete longitudinal tear occurring at the time of the initial injury is considered to be rare. Completed tears are considered to occur from repeated injuries.

The medial meniscus sustains its initial tear most often in the posterior pole, and a longitudinal tear usually occurs in the posterior third of the meniscus. If the tear extends anteriorly past the collateral ligament, it bunches up between the condyles and can cause the joint to "lock." Extensive tearing may result in a fragment protruding into the center of the joint with *no* resultant locking of the knee joint.

The effusion of a meniscus tear results from injury to the synovium. A large effusion can cause pain and limited motion.

Other causative factors are implicated, including constitutional inadequacy, ligamentous laxity, faulty work or sports habits, excessive knee valgus or varus, violent sports, or a previous injury inadequately treated.

Diagnosis of Medial Meniscus Injury. The diagnosis of medial meniscus injury is made by a careful evaluation of the history, which describes the mechanism of the injury. This may be furnished by the injured person

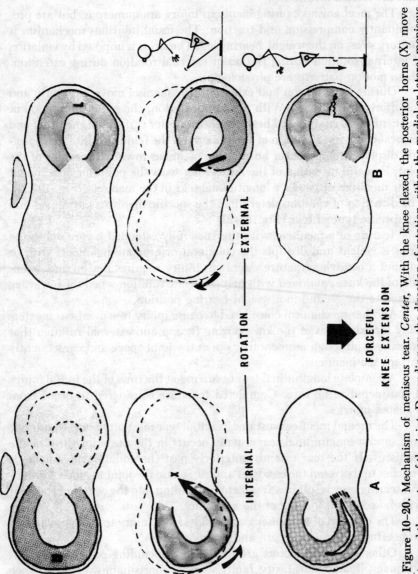

Figure 10–20. Mechanism of meniscus tear. *Center*, With the knee flexed, the posterior horns (X) move toward the center of the joint. Depending on the direction of rotation, either the medial or lateral meniscus so moves. *Below*, As the person extends the knee, the meniscus can sustain a tear.

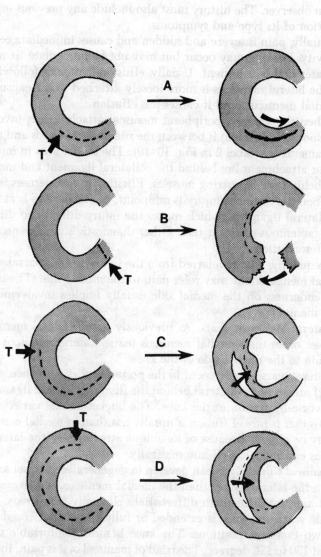

Figure 10–21. Types of meniscus tear (T). *A,* Tear in posterior third of meniscus. Because of the elasticity of the cartilage, the meniscus springs back into its normal position. *B,* Posterior tear with avulsion causes the anterior portion of the meniscus to bunch. *C,* Longitudinal partial tear may cause bunching of inner third of meniscus with resultant locking. *D,* Complete longitudinal bucket handle tear. The central portion of the meniscus moves into the center of the joint away from the condyles and thus the joint does not lock.

or by an observer. The history must also include any previous injury, with description of its type and symptoms.

Usually, pain is severe and sudden and causes immediate cessation of all activity. Locking may occur but may not be immediate. It may occur hours later and be transient. Usually, effusion is present following an injury. The lateral meniscus is more loosely attached to the capsule than is the medial meniscus, and it causes less effusion.

When the anterior peripheral meniscal attachment is involved, the tender spot (trigger area) is between the site of the fat pads and the collateral ligaments (see label 2 in Fig. 10–19). The tender area in injury to the posterior attachment lies behind the collateral ligament and may be concealed behind the hamstring muscles. Eliciting the tenderness is difficult.

When the meniscus injury is midpoint, the tenderness is at the site of the collateral ligament, which makes the injury difficult to differentiate from ligamentous strain or tear. Other diagnostic signs are necessary for this differentiation.

Frequently, pain is referred from the injury to the contralateral side. A lateral meniscal tear may refer pain to the medial side. The converse is rare. Tenderness on the medial side usually implies involvement of the medial meniscus.

Lateral Meniscus Tears. As previously stated, lateral meniscus tears occur less often than medial meniscus tears. Lateral meniscus tears can refer pain to the medial side of the knee.

Lateral tears usually occur in the posterior portion; hence, the tender (trigger) area is posterolateral behind the fibular collateral ligament and is made worse by *flexion* of the knee. The bucket-handle variety of lateral meniscus tear is rare. Effusion is usually less than in medial meniscus tear and may be absent. Because of its unique attachments, the lateral muscle meniscus can reduce itself automatically.

Meniscal Cysts. Cysts can develop in degenerated menisci and usually occur in the lateral rather than the medial meniscus. These cysts may be painful and are difficult to differentiate clinically from tears. The cysts protrude when the knee is extended or fully flexed, but recede between these two excessive positions. The knee is most comfortable when it is flexed at 130 to 120 degrees. The triad of meniscal cyst is pain, interference with joint motion (full extension and full flexion), and a palpable tender tumor that recedes with slight flexion.

Diagnostic Meniscus Signs. Meniscus signs (Fig. 10–22), also termed signs of internal derangement of the knee, are numerous and unfortunately are confusing in that many are termed by the proper name of the physician who originally described the specific maneuver (e.g., McMurray's sign, Apley test). All of these tests (which are described in *Knee Pain and Disability*[1]) attempt to reproduce the symptoms or confirm the mechanical impairment by flexing or extending the knee with various degrees of rotation.

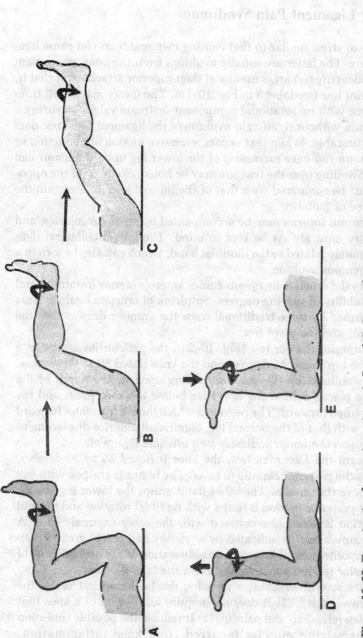

Figure 10–22. Examination for meniscus signs. A, B, and C, Stages of McMurray test. With the patient supine, the knee is fully flexed. The leg is internally rotated (lateral meniscus test) or externally rotated (medial meniscus test), and then the knee is fully extended. In a meniscus lesion, a painful click is elicited which is most significant in the first phase of extension. D and E, Apley test. With a patient prone and leg rotated upward, pain upon traction (D) implies a capsular ligamentous lesion, while pain upon downward pressure (E) indicates a meniscus lesion.

Collateral Ligament Pain Syndromes

A force or stress similar to that causing meniscal tears can cause ligamentous tears. The latter are usually avulsions from the bony attachment with the tender (trigger) areas usually at their superior attachment, that is, above the joint line (see label 3 in Fig. 10–19). The injury may result from a lateral force with no rotational component (extreme valgus or varus).

In a strain without significant avulsion of the ligament, the knee does not become unstable. When tear occurs, excessive motion of abduction or adduction *(with full knee extension)* of the lower leg upon the femur can be elicited. Swelling over the tear site may be noted. Stability of the opposite knee must be compared with that of the injured knee to ascertain the latter's degree of mobility.

Ligamentous injuries may be accompanied by meniscus injuries, and this possibility must always be kept in mind. The lateral collateral ligament is intimately related to the iliotibial band, which can also be torn in a severe ligamentous avulsion.

Injury to the cruciate ligaments causes anteroposterior instability and rotatory instability of varying degrees. Suspicion of cruciate ligament tear can be confirmed by three traditional tests: the anterior drawer test, the Lachman test, and the pivot test.

In the anterior drawer test (Fig. 10–23), the patient lies supine on a table with the hip flexed 45 degrees and the knee flexed 80 to 90 degrees. The foot is stabilized by the examiner sitting upon it, the hands of the examiner are placed behind the lower leg below the knee joint, and the lower leg is pulled forward. The degree to which the leg translates forward is compared with that of the normal leg. Significant anterior displacement of the tibia upon the femur indicates torn cruciate ligaments.

To perform the Lachman test, the knee is flexed 15 to 20 degrees, which allows the posterior capsule to relax. The femur is grasped with one hand just above the patella. The other hand grasps the lower leg (tibia). The anterior-posterior motion is tested with no tibial rotation and the end point of motion is noted as compared with the other (normal) knee. A "positive Lachman test" is indicated by a "mushy end point" greater than that of the opposite normal knee and an obliteration of the normal slope of the infrapatellar tendon when viewed from the lateral aspect.

There are several advantages to doing the Lachman test versus the standard "drawer test." (1) It does not require full flexion of a knee that may be swollen (effusion) and painful. (2) It reduces the possible confusion of spastic hamstring muscles incurred from knee inflammation. (3) It minimizes the confusing effect if there is a concommitent posterior meniscus tear that "blocks" full movement of the knee joint. (4) The Lachman test is considered to be more sensitive and accurate than the drawer test in diagnosing an anterior cruciate tear. The major disad-

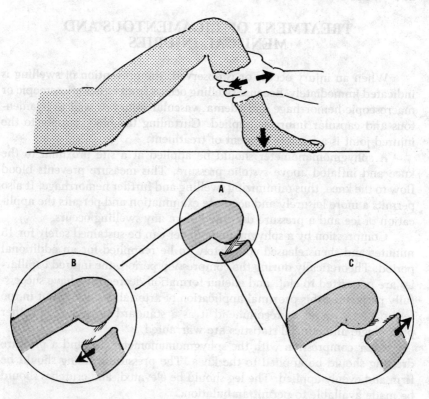

Figure 10-23. Drawer sign for cruciate ligament tears. *Top*, Position of patient's leg and examiner's hand. The foot is fixed upon the table, and the lower leg is moved horizontally, proximally, and distally *(arrows)*. *Bottom*, *A*, Intact cruciate ligaments. *B*, Posterior tibial movement in tear of posterior cruciate ligament. *C*, Excessive anterior tibial motion in tear of anterior cruciate ligament. If both cruciate ligaments are torn, there is excessive motion in both directions.

vantage of performing the Lachman test is that the examiner must have relatively large hands to perform the test adequately.

The pivot test is performed with the patient lying supine on a table with the hip and the knee flexed to 90 degrees and the foot in the air. The ankle is grasped, and the tibia is internally rotated. The examiner's other hand imposes a valgus force stressing the knee upon the femur. Any subluxation of the joint denotes a torn ligament.

TREATMENT OF LIGAMENTOUS AND MENISCAL INJURIES

When an injury occurs or is observed, the prevention of swelling is indicated immediately. Because swelling occurs from either microscopic or macroscopic hemorrhage and edema, vascular injury as well as ligamentous and capsular injury is implied. Curtailing the blood supply to the injured joint is a feasible element of treatment.

A sphygmomanometer should be applied at a site proximal to the knee and inflated above systolic pressure. This measure prevents blood flow to the knee, thus minimizing swelling and further hemorrhage. It also permits a more leisurely and accurate examination and permits the application of ice and a pressure dressing before any swelling occurs.

Compression by a sphygmomanometer can be sustained safely for 15 minutes and then released completely, to be reapplied for an additional period. Theoretically during the compression period, the injured capillaries are permitted to clot, and edema formation decreases. I have successfully performed this proximal application of arterial occlusion, but in too few patients as yet to recommend it as a standard procedure. Further physiologic and clinical statistics are warranted.

After compression with the sphygmomanometer, ice and a pressure dressing should be applied to the knee. The pressure dressing should be firm and evenly applied. The leg should be elevated, and crutches should be made available to permit ambulation.

If a sphygmomanometer is not available, or if there is no one present with knowledge of its use, elevation of the leg, ice packing, and pressure dressings are indicated.

The application of a tourniquet or a sphygmomanometer after the occurrence of edema is not indicated and may actually be damaging. No constriction to the circulation should be applied unless there is a knowledgeable person present and who will remain with the injured person until the pressure is released.

Ligamentous Injuries

Ligamentous strains or minor tears actually heal well in time, but active exercises must be instituted early. Initially, isometric exercises, which avoid joint motion but increase muscle strength and endurance, are prescribed (Fig. 10–24).

Isometric exercises are gradually followed by kinetic (isotonic) exercises, and then by progressive resistive exercises (Fig. 10–25).

If significant effusion occurs in the joint, aspiration should be performed early, and repeated later if effusion recurs (Fig. 10–26). Persistent

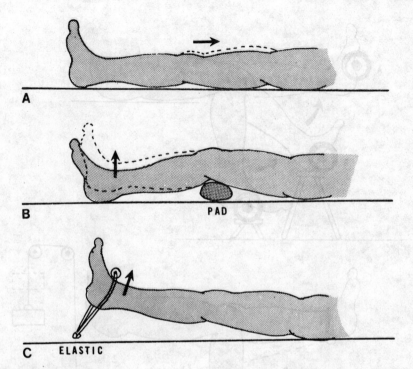

Figure 10-24. Types of quadriceps exercises. *A*, Quad setting. Knee is fully extended, muscle shortens with *no* joint motion, and patella ascends. *B*, With a 3-inch pad under the knee, the quadriceps function is enhanced, and joint motion is minimal. *C*, Straight leg raising against resistance (resisted isometric exercise) further strengthens quadriceps without joint motion.

effusion, especially if it is great and creates pressure and hemorrhage, causes mechanical and ischemic damage to the already damaged, overstretched, and torn tissue.

Meniscal Injuries

Meniscal tears in the inner two thirds of the meniscus (the avascular zone) do not heal, whereas tears in the outer vascular zone that are reduced can heal.

If the displaced meniscus can be reduced and if the leg can be immobilized for at least three weeks, recovery is possible. There are no specific indications for surgery, but the decision for surgery depends on the surgeon's experience and expertise. As a rule, recurrent locking, sustained disability, or intractable pain suggests a need for arthrography and surgery.

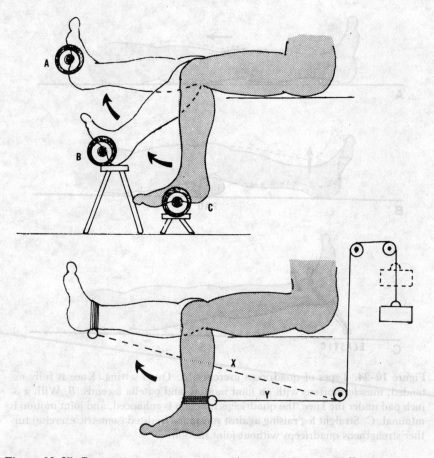

Figure 10-25. Progressive resistive exercises (isotonic). *Top, A,* Full extension is attempted. *B,* With stool to support weighted foot, only the last degrees of extension are exercised. This exercise is most desired and strengthens the vastus medialis. *C,* Resistance is minimal and there is ligamentous strain on the knee when the leg is fully dependent. *Below, Bottom,* Pulley exercises. Maximum resistance occurs in first 45 degrees of extension, and none occurs at full extension.

Reduction must be initiated within 24 hours of locking. Reduction technique requires manual longitudinal traction with simultaneous rotation in both directions and lateral mobilization (into varus and valgus). The knee is placed in full flexion and forcefully rotated—internally in a medial meniscus tear and externally in a lateral meniscus tear. Then, forceful kicking by the patient places the knee in full extension. If the procedure is unsuccessful, repeated manipulations or forceful reduction attempts should not be considered.

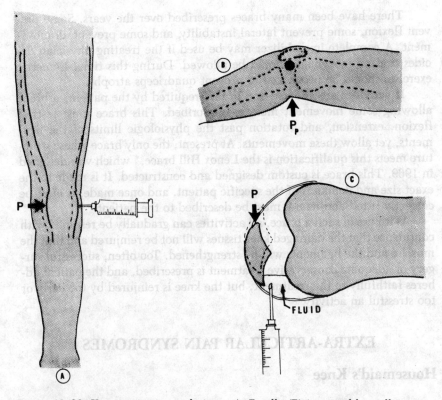

Figure 10-26. Knee aspiration technique. *A*, Patella (P) is moved laterally to increase injectable space between the patella and the femoral condyle. *B*, Lateral view of site of injection (black dot). Pressure upon popliteal space (P_P) brings fluid toward needle tip. *C*, In lateral position, gravity localizes fluid to permit easier withdrawal.

After successful reduction and adequate immobilization, restoration exercises should be undertaken. The major purpose of the exercise program should be to strengthen the quadriceps mechanism.

After an injury to the knee, whether it is a strained or torn ligament or a torn meniscus treated conservatively or surgically, the knee is unstable. The exercise program of strengthening the quadriceps also gradually strengthens the ligaments. Sufficient time is required to restabilize the knee so that it can function safely and comfortably.

Four weeks are needed for the ligaments to regain 60 percent of their tensile strength, and six weeks are needed to regain 80 percent. The knee must be protected if activities, especially athletic activities, are to be resumed during this time. A brace is mandatory to allow healing, prevent further injury, and permit adequate function.

There have been many braces prescribed over the years. Some prevent flexion, some prevent lateral instability, and some prevent all movement. A complete immobilizer may be used if the treating physician decides that *no* movement should be allowed. During this time, isometric exercises should be prescribed to prevent quadriceps atrophy.

If activities are allowed and in fact required by the patient, a brace allowing some movement must be prescribed. This brace must restrict flexion, extension, and rotation past the physiologic limits of the ligaments, yet allow these movements. At present, the only brace whose structure meets this qualification is the Lenox Hill brace,* which was designed in 1969. This brace is custom designed and constructed. It is made to the exact size and demands of the specific patient, and once made, it must be correctly fitted and its use must be described to the patient.

With use of such a brace, all activities can gradually be resumed with confidence that the damaged knee tissues will not be reinjured and that the muscles and the ligaments will be strengthened. Too often, successful surgery or adequate conservative treatment is prescribed, and the patient adheres faithfully to the treatment, but the knee is reinjured by too early or too stressful an activity.

EXTRA-ARTICULAR PAIN SYNDROMES

Housemaid's Knee

There is a bursa located beneath the skin and between the patella and the tibial tubercle that is subject to mechanical pressure during kneeling. Essentially, there are two bursae in this region: one over the lower half of the patella and one over the superior half of the tibial tubercle (Fig. 10–27).

When the bursae are inflamed, aching or pain occurs in response to local pressure or upon kneeling. Swelling is usually noted, and serous fluid can be aspirated.

Pes Anserinus Bursitis

The location of the pes anserinus bursa is superficial to the medial collateral ligament at the upper medial aspect of the tibia under the conjoined tendon of the sartorius, semimembranous, and gracilis muscles. Pain is increased by forceful extension of the knee or by resisted contraction of the hamstrings in knee flexion.

---

*Lenox Hill Brace, Long Island, NY.

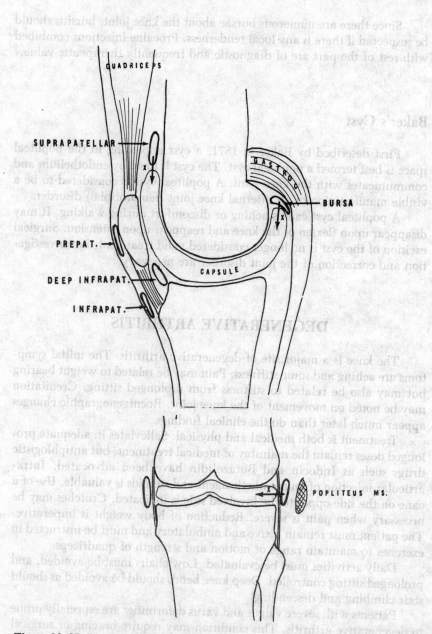

Figure 10-27. Bursae about the knee. The suprapatellar bursa may communicate with the knee capsule. X indicates other bursae that may communicate with joint space.

Since there are numerous bursae about the knee joint, bursitis should be suspected if there is any local tenderness. Procaine injections combined with rest of the part are of diagnostic and frequently therapeutic value.

Baker's Cyst

First described by Baker in 1871, a cyst appearing in the popliteal space is best termed a **popliteal cyst**. The cyst is lined by endothelium and communicates with the knee joint. A popliteal cyst is considered to be a visible manifestation of an internal knee joint (femorotibial) disorder.

A popliteal cyst causes aching or discomfort during walking. It may disappear upon flexion of the knee and reappear upon extension. Surgical excision of the cyst is no longer considered valid treatment, but investigation and correction of the joint disorder are necessary.

DEGENERATIVE ARTHRITIS

The knee is a major site of degenerative arthritis. The initial symptoms are aching and some stiffness. Pain may be related to weight bearing but may also be related to stiffness from prolonged sitting. Crepitation may be noted on movement of the knee joint. Roentgenographic changes appear much later than do the clinical findings.

Treatment is both medical and physical. Salicylates in adequate prolonged doses remain the mainstay of medical treatment, but antiphlogistic drugs such as Indocin and Butazolidin have been advocated. Intraarticular injection of an anesthetic agent and steroids is valuable. Use of a cane on the side opposite the involved side is indicated. Crutches may be necessary when pain is severe. Reduction of body weight is imperative. The patient must remain active and ambulatory and must be instructed in exercises to maintain range of motion and strength of quadriceps.

Daily activities must be evaluated. Low chairs must be avoided, and prolonged sitting controlled. Deep knee bends should be avoided as should stair climbing and descending.

Patients with severe valgus and varus deformities are especially prone to degenerative arthritis. This condition may require bracing or surgical intervention. Surgical procedures for relief of degenerative arthritis pain are beyond the scope of this book.

Rheumatoid arthritis is a systemic disease in which the knee is usually involved. The initial phase of joint disease is soft tissue involvement with effusion, heat, redness, pain, and limited motion. Diagnosis is made by history, general physical findings, laboratory test confirmation, synovial fluid examination, and ultimately roentgenographic findings.

Treatment has been well documented elsewhere,[1] but I wish to emphasize that quadriceps exercises *must* be started early to avoid atrophy. Flexion contraction must be avoided, and oral medication and intra-articular injection of steroids after aspiration are valuable (see Fig. 10-26).

Treatment may ultimately involve surgical intervention.

Chondromalacia Patellae

Painful disorder of the patellofemoral articulation has come to be termed chondromalacia patellae, implying that any joint pain of the patellofemoral joint is degenerative disease of the patellar cartilage. Knee pain in the patellofemoral joint has many other causes and thus the term chondromalacia patellae should not be loosely nor universally used.

Pain in the patellofemoral joint is common; in fact, it is considered to be the most prevalent type of knee pain. There are tremendous forces exerted upon this joint in everyday activities, and it has been estimated that by age 30, everyone has some degeneration of the patellar cartilage. Some crepitation on movement is noted in almost everyone past that age.

The patellofemoral joint is also subjected to lateral or medial deviation from its arcing course upon the femoral condyles. The patella may "slip," sublux, or actually dislocate. This clinical condition is not truly chondromalacia of the patella although it may be painful and disabling.

Patellofemoral arthralgia, a better term, is pain felt in the kneecap region when the knee functions under load pressures. Pain is felt characteristically upon going up or down steps: usually the latter. Pain may be felt after prolonged sitting and then upon arising and may be relieved by extending the knee.

The painful condition usually occurs after one of the following types of trauma: (1) sudden direct violent impact such as can occur in automobile accidents when the knee strikes the dashboard and (2) repetitive minor trauma due to alterations from mechanical malalignment.

In direct trauma, the cause and effect relationship is obtained from the history. The malalignment is determined from examination revealing (1) genu valgum, (2) genu varum, (3) tibial torsion, (4) patella alba, or (5) excessive Q angle from other causes.

Pain and impairment may occur without cartilage degeneration. Only when there is degeneration is the term chondromalacia patellae justified. This degeneration is currently verifiable with arthroscopic examination. When severe, it can be ascertained by x-ray studies.

There are four stages of chondromalacia based on the degree of cartilage damage:

Stage I: Softening of the cartilage
Stage II: Blister formation
Stage III: Ulceration
Stage IV: Crater formation and eburnation

These stages are well illustrated by Shahriaree.[3]

Symptoms. Patients are usually young active people. Patients complain of pain near the anterior aspect of the knee, felt after exercises or after climbing or descending stairs. "Stiffness" is noted after prolonged sitting. The knee may "give way." Crepitus is felt on movement of the joint. Tenderness may be felt under the medial or the lateral edge of the patella. Pain can be reproduced by contraction of the quadriceps and simultaneous resistance of the patellar movement.

X-ray studies may verify a cause of the condition, but the diagnosis is primarily clinical.

Treatment. Nonsurgical treatment should be administered for a long period before surgical intervention is considered. Changes in lifestyle are required: deep knee bends and stairs should be avoided, and sports that require movements that initiate the pain should be avoided.

Isometric quadriceps exercises with gradual increase of resistance and with 20 repetitions performed daily are effective. Isotonic exercises performed against resistance in the last 20 degrees of extension should also be included in the regimen.

When there is evidence of lateral or medial deviation of the patella, a knee brace with the patella site cut out should be worn. This brace guides the movement of the patella and keeps it in the femoral condylar path.

After an acute episode of pain, treatment may include rest, application of ice, oral anti-inflammatory medication, or even a knee splint or cast, depending on the severity of the pain.

As symptoms of chondromalacia patellae may coexist with other knee disorders, such as cruciate ligament laxity or meniscus injury, these other disorders must be ruled out or treated.

When conservative measures of treatment fail, surgery may be needed. Surgical procedures include shaving the patellar cartilage, shortening or lengthening the parapatellar ligaments, and even the Maquet procedure.

The Maquet procedure involves anterior displacement of the tibial attachment of the patellar tendon a distance of 2.5 cm (Fig. 10–28). This displacement is accomplished by performing an osteotomy of the tibial tubercle and inserting a bone block. By changing the vector angle of the patellar tendon, the force of the patella against the femoral condyles is decreased by as much as 50 percent. With this new position of the patella, the upper facet of the patella is applied against the femur during knee flexion sooner than is the middle facet. Since the upper facet is larger,

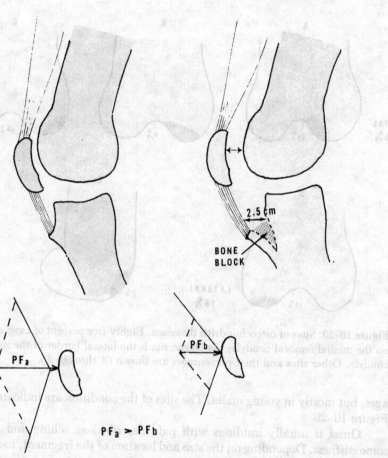

Figure 10–28. Anterior advancement of patellar tendon. *Left*, Normal patellofemoral relationship. *Right*, A 2.5-cm advancement of the tibial tubercle attachment of the patellar tendon. The patella is more distant from the femoral condyle. The parallelograms reveal the patellar force (PF) to be less in the right figure.

there is a relative decrease of surface weight-bearing area.

The lateral or medial deviation (Q angle) can be changed.[4] This procedure is applicable for symptoms of chondromalacia or for recurrent patellar dislocation.

OSTEOCHONDRITIS DISSECANS

Osteochondritis dissecans, a painful condition, is caused by a fragment of cartilage becoming loose in the joint. It occurs in persons of all

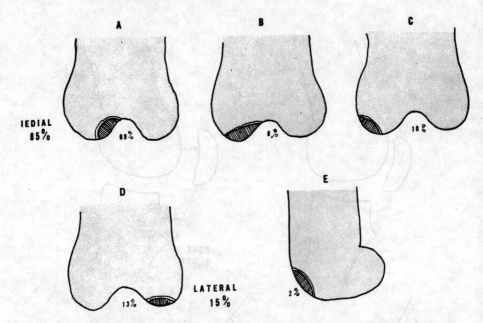

Figure 10–29. Sites of osteochondritis dissecans. Eighty-five percent of cases occur on the medial femoral condyle; the classic site is the lateral border of the medial condyle. Other sites and their percentages are shown (A through E).

ages, but mostly in young males. The sites of the condition are indicated in Figure 10–29.

Onset is usually insidious with pain described as aching and with some stiffness. Depending on the size and location of the fragment, locking can occur.

Diagnosis is confirmed by roentgenography. If time and periodic immobilization do not afford relief, surgical excision is indicated.

OSGOOD-SCHLATTER DISEASE

Osteochondritis of the tibial tubercle is called **Osgood-Schlatter disease.** It is observed mostly in adolescent boys. The patient complains of pain and tenderness over the tibial tubercle, and these symptoms are aggravated by kneeling. Swelling and tenderness over the tibial tubercle are noted clinically. Etiologic factors have not been determined, and roentgenography usually yields negative results early in the disease.

Treatment consists of avoiding excessive activities such as kneeling, squatting, or jumping. In severe cases, cylindric casts can be used to prevent the patient from squatting or kneeling.

Larsen-Johannson disease is a variant of Osgood-Schlatter disease in which the osteochondritis affects the inferior pole of the patella. Symptoms and treatment are identical to those of Osgood-Schlatter disease, and diagnosis is confirmed by roentgenography.

REFERENCES

1. Cailliet, R: Knee Pain and Disability, ed 2. FA Davis, Philadelphia, 1983.
2. Steindler, A: Lectures on the Interpretation of Pain in Orthopedic Practice. Charles C Thomas, Springfield, IL, 1959, Lecture XV, pp 555–605.
3. Shahriaree, H: Chondromalacia. Contemporary Orthopedics 11:27–39, Nov 1985.
4. Insall, J, Falvo, KA, and Wise DW: Chondromalacia patellae. J Bone Joint Surg [Am] 58:1–8, 1976.

BIBLIOGRAPHY

Basmajian, JV: Grant's Method of Anatomy. Williams & Wilkins, Baltimore, 1971.

Basmajian, JV, and Lovejoy, JF: Function of the popliteus muscle in man. J Bone Joint Surg [Am] 53:557, 1971.

Brantigan, OC, and Voshell, AF: Tibial collateral ligament: Its function, its bursae, and its relation to the medial meniscus. J Bone Joint Surg 25:1, 1943.

Crooks, LM: Chondromalacia patellae. J Bone Joint Surg 49-B:495–501, 1967.

DeLorme, TL: Restoration of muscle power by heavy resistance exercises. J Bone Joint Surg [Am] 27:645, 1945.

DePalma, AF: Diseases of the Knee: Management and Surgery. JB Lippincott, Philadelphia, 1954.

Goodfellow, J, and Woods, C: Patello-femoral joint mechanics and pathology, 1. Functional anatomy of the patello-femoral joint. J Bone Joint Surg 59-B:287–291, 1976.

Goodfellow, J, and Woods, C: Patello-femoral joint mechanics and pathology, 2. Chondromalacia patellae. J Bone Joint Surg 58-B:291–299, 1976.

Helfet, AJ: Mechanism of derangements of the medial semilunar cartilage and their management. J Bone Joint Surg [Br] 41:319, 1959.

Helfet, AJ: Function of cruciate ligaments of knee-joint. Lancet 1:665, 1948.

Jack, EA: Posterior peripheral detachment of the lateral cartilage. J Bone Joint Surg [Br] 35:396, 1953.

Kuhns, JG: Changes in elastic adipose tissue. J Bone Joint Surg [Am] 31:541, 1949.

Last, RJ: Some anatomical details of the knee joint. J Bone Joint Surg [Br] 30:683, 1948.

Lieb, FJ, and Perry, J: Quadriceps function. An electromyographic study under isometric conditions. J Bone Joint Surg [Am] 53:749, 1971.

MacConall, MA, Barnett, CH, and Davies, DV: Synovial Joints: Their Structure and Mechanics. Charles C Thomas, Springfield, IL, 1961.

Maquet, P: Un traitement biomecanique de l'arthrose femero-patellaire. L'Avancement dei tendon ratalein. Rev Rhumat 30:779, 1963.

Marshal, JL, Girgis, FG, and Zelko, RR: The biceps femoris tendon and its functional significance. J Bone Joint Surg [Am] 54:1444, 1972.

McMurray, TP: The semilunar cartilage. Br J Surg 29:407, 1942.

Murray, JUG: The Maquet principle: Its application in severe chondromalacia patellar, patellofemoral and global knee osteoarthritis. Orthopedic Review 8:29–36, 1976.

Ricklin, P, Ruttmann, A, and Del Buono, MS: Meniscus Lesions: Practical Problems of Clinical Diagnosis, Arthrography, and Therapy. Grune & Stratton, New York, 1971.

Shahriaree, H: Chondromalacia. Contemporary Orthopedics 11:27-39, Nov 1985.

Slocum, DB, and Larson, RL: Rotatory instability of the knee. J Bone Joint Surg [Am] 50:211, 1968.

Thorek, SL: Orthopaedics: Principles and Their Application, ed 2. JB Lippincott, Philadelphia, 1967.

Wolfe, RD, and Colloff, B: Popliteal cysts. J Bone Joint Surg 54-A:1057-1063, 1972.

CHAPTER 11

Foot and Ankle Pain

The foot presents a unique segment of human anatomy in that all its parts are accessible to visual examination, direct palpation, and mechanical evaluation. The patient directs the attention of the examiner to the painful area or the disabling function, and the examination should lead to a specific diagnosis. In evaluation of foot disorders, complete knowledge of the functional anatomy is necessary to an even greater degree than in evaluation of other portions of the musculoskeletal system. Meaningful treatment logically follows meaningful evaluation.

FUNCTIONAL ANATOMY

The foot is an intricate structure composed of 26 articulating bones constructed to bear full body weight and to transport the human body over varying types of terrain. The foot can be divided into three functional units: anterior, middle, and posterior (Fig. 11–1).

The posterior segment lies directly under the tibia and supports the superincumbent body. The talus in this segment is the mechanical keystone of the weight-bearing foot, articulating superiorly into the ankle mortice and inferiorly to the ground by contact with the calcaneus.

The talus has a body, a neck, and a head (Fig. 11–2). The superior portion and both sides of the body articulate with the tibia and fibula, which unite to form the ankle mortice. The tibia contacts the entire superior aspect of the talus and is weight bearing. Its medial malleolus extends one third of the way down the medial aspect of the talus. The fibular malleolus covers the entire lateral aspect of the body of the talus. Within

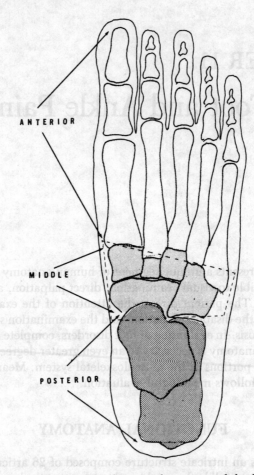

Figure 11-1. The three functional segments of the foot.

the mortice, the talus functions as a hinge joint and permits dorsal and plantar motion of the foot at the ankle (Fig. 11–3).

Viewed from above, the talus is wedge-shaped with a wider anterior portion. As the ankle dorsiflexes, this wider portion moves between the two malleoli of the mortice and wedges there, separating the ankle mortice. No lateral or rotatory motion of the talus within the mortice is possible in this dorsiflexed position. During plantar flexion of the ankle, the narrower posterior portion of the talus moves between the malleoli and allows mobility within the joint. This unstable position places added burden on the ligaments, which must bear the brunt of stabilizing the joint.

The talus moves in the mortice at an 18-degree angle because its axis of rotation is oblique to the coronal plane. The axis passes laterally through

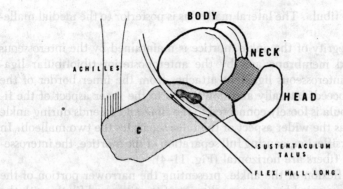

Figure 11-2. Talus. Comprising a body, neck, and head, the talus has two articulating surfaces that fit into the ankle mortice. The head articulates with the bones of the middle segment, and the entire talus sits upon the calcaneus (C).

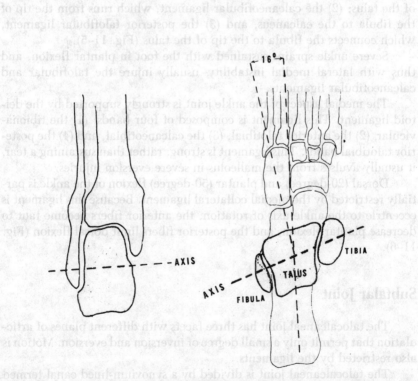

Figure 11-3. Ankle mortice and talus relationship. *Left,* The axis of rotation passes through the fibula and below the tip of the tibia. *Right,* Viewed from above, the medial malleolus is anterior to the fibula, forming a 16-degree toe-out stance and gait. The talus also is broader anteriorly than posteriorly.

the tip of the fibula. The lateral malleolus is posterior to the medial malleolus.

The integrity of the ankle mortice is maintained by the interosseous ligament and membrane and by the anteroposterior tibiofibular ligaments. The interosseous ligament attaches from the inner border of the tibia and proceeds laterally and downward to the inner aspect of the fibula. The fibula is loosely connected to the tibia and ascends during ankle dorsiflexion as the wider aspect of the talus separates the two malleoli. In full ankle dorsiflexion, causing full separation of the mortice, the interosseous ligament fibers are horizontal (Fig. 11–4).

Plantar flexion of the ankle, presenting the narrower portion of the talus, is accompanied by reapproachment of the tibia and fibula with the interosseous fibers resuming their oblique direction.

The ankle joint receives strong support from the collateral ligaments. The lateral collateral ligament is composed of three bands: (1) the anterior talofibular ligament, which attaches from the tip of the fibula to the neck of the talus; (2) the calcaneofibular ligament, which runs from the tip of the fibula to the calcaneus, and (3) the posterior talofibular ligament, which connects the fibula to the tip of the talus (Fig. 11–5).

Severe ankle sprains sustained with the foot in plantar flexion, and thus with lateral medial instability, usually injure the talofibular and calcaneofibular ligaments.

The medial aspect of the ankle joint is strongly supported by the deltoid ligament. This ligament is composed of four bands: (1) the tibionavicular, (2) the anterior talotibial, (3) the calcaneotibial, and (4) the posterior talotibial bands. This ligament is strong; rather than sustaining a tear, it usually avulses from the malleolus in severe eversion injuries.

Dorsal (20-degree) and plantar (50-degree) flexion of the ankle is partially restricted by the medial collateral ligament. Because this ligament is eccentric to the ankle axis of rotation, the anterior fibers become taut to decrease plantar flexion, and the posterior fibers limit dorsal flexion (Fig. 11–6).

Subtalar Joint

The talocalcaneal joint has three facets with different planes of articulation that permit only a small degree of inversion and eversion. Motion is also restricted by the ligaments.

The talocalcaneal joint is divided by a synovium-lined canal termed the tarsal canal. This canal is funnel-shaped with the wide portion at the lateral end located just slightly below and anterior to the lateral malleous. The opening is increased by inverting the foot in plantar flexion. The canal

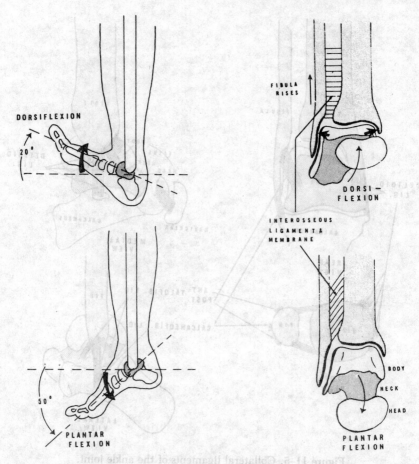

Figure 11-4. Motion of talus within mortice: dorsiflexion and plantar flexion. As the ankle dorsiflexes, the wider anterior portion of the talus wedges in the mortice, and the fibula rises, causing the interosseous ligament to become horizontal. When the mortice has been fully expanded, further dorsiflexion is prevented. On plantar flexion, the narrow portion of the talus presents itself, and the fibula descends. Here the fibers resume their oblique direction, and the mortice decreases its width.

proceeds medially and posteriorly to its medial opening, located just behind and above the sustentaculum tali. The angle that the canal forms with the anteroposterior axis of the foot is approximately 45 degrees. The canal is of sufficient width, thus making intra-articular injections relatively simple (Fig. 11-7).

A firm talocalcaneal ligament binds the two bones. This ligament runs the length of the tarsal canal, but is more discrete at the fibular end to

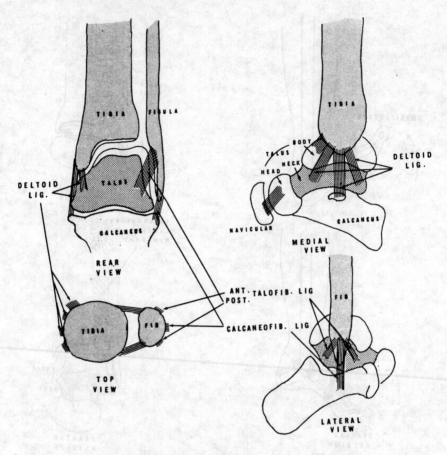

Figure 11–5. Collateral ligaments of the ankle joint.

permit some rotation about this point. The talocalcaneal ligament, because of its direction within the tarsal canal, becomes taut during inversion and slackens during eversion. This action makes the inverted (supinated) foot more stable.

Movement of the talocalcaneal joint can be isolated and tested by full dorsiflexion of the ankle, a position that locks the talus in the mortice. Then, by moving the calcaneus, the examiner can document the extent of joint motion and localize pain originating from the joint.

The axis about which the calcaneus rotates upon the talus forms a 45-degree angle to the floor and a 16-degree angle in a medial direction from a line drawn through the second metatarsal (Fig. 11–8).

This axis is termed the subtalar axis. Three movements occur about this axis:

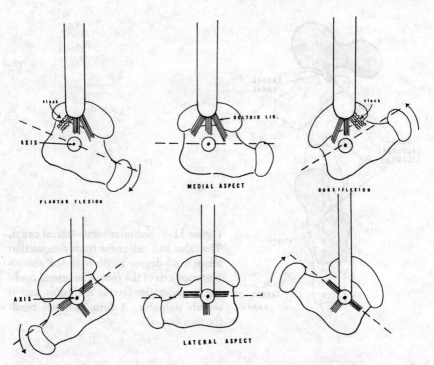

Figure 11-6. Relationship of medial and lateral collateral ligaments to the axis of ankle motion. The transverse axis of the ankle mortice transects the center of attachment of the lateral ligaments at the tip of the fibula. Plantar flexion or dorsiflexion of the ankle does not change the length of the ligaments. The axis is eccentric to the medial deltoid ligament. During plantar flexion, the posterior strands become slack, and the anterior fibers become taut. The opposite occurs during dorsiflexion.

1. **Inversion** about the longitudinal axis. This movement consists of elevation of the medial border as the lateral border of the foot depresses.
2. **Abduction and adduction.** These movements are outward rotation and inward rotation, respectively, about an axis drawn vertically through the tibia.
3. **Dorsal and plantar flexion** about the transverse axis. This last motion is significantly more restricted than the similar motion of the talus within the ankle mortice.

Supination of the foot is a combined motion of the three subtalar motions, namely inversion, adduction, and plantar flexion. **Pronation** combines eversion, abduction, and dorsiflexion.

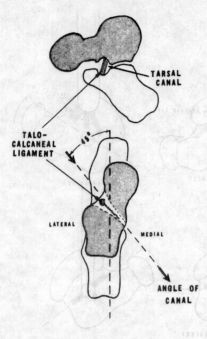

Figure 11-7. Subtalar joint—tarsal canal. The talus and calcaneus from a canal that forms a 45-degree angle with the antero-posterior axis of the foot. The lateral opening is under the lateral malleolus and is readily palpable. A firm ligament bends the two bones.

Transverse Tarsal Joint

The transverse tarsal joint combines essentially two joints in an axial alignment: the talonavicular and the calcaneocuboid joints (Fig. 11-9). The transverse tarsal joint has been alternately termed Chopart's joint, the midtarsal joint, and because it is the site of an elective amputation, the surgeon's tarsal joint.

The planes of the talonavicular and the calcaneocuboid joints differ. The rounded head of the talus fitting into the concavity of the navicular bone permits rotation about this axis during pronation and supination of the forefoot and some gliding motion during inversion and eversion of the foot.

The calcaneocuboid joint has limited range of motion that permits a slight degree of abduction and adduction of the forefoot. When the axis of the talonavicular joint parallels the axis of the calcaneocuboid joint, as it does in the pronated foot, there ensues some motion that results in instability. In the supinated foot, both axes diverge, decreasing joint mobility and thus creating a more stable foot. This factor plus the tautness of the talo-calcaneal ligament explains the stability of the supinated foot. Also, in supination, all articular surfaces are closely packed, that is, in complete opposition, and thus are more stable in terms of engineering and less dependent on ligamentous support.

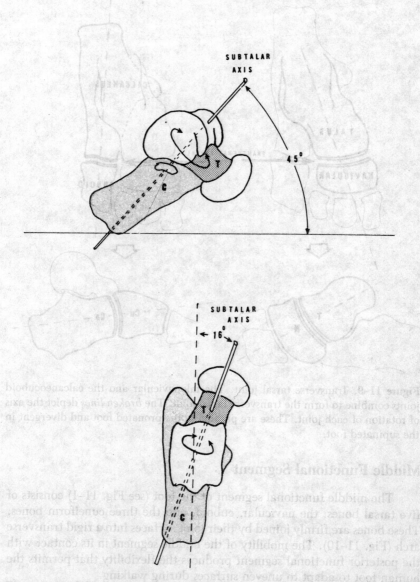

Figure 11–8. Subtalar axis. Movement about the subtalar axis consists of supination and pronation of the foot. The axis forms a 45-degree angle with the ground and a 16-degree angle medial to a longitudinal line drawn through the second metatarsal.

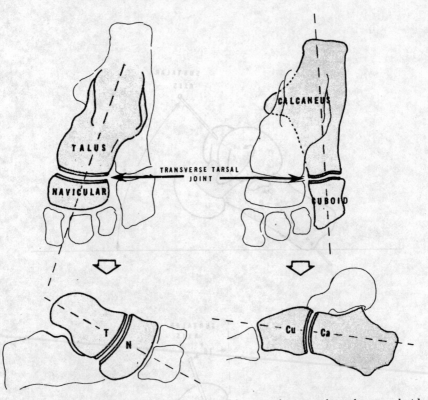

Figure 11–9. Transverse tarsal joint. The talonavicular ana the calcaneocuboid joints combine to form the transverse tarsal joint. The *broken lines* depict the axis of rotation of each joint. These are parallel in the pronated foot and divergent in the supinated foot.

Middle Functional Segment

The middle functional segment of the foot (see Fig. 11–1) consists of five tarsal bones: the navicular, cuboid, and the three cuneiform bones. These bones are firmly joined by their joint surfaces into a rigid transverse arch (Fig. 11–10). The mobility of the middle segment in its contact with the posterior functional segment produces the flexibility that permits the human foot to adapt to uneven surfaces during walking.

The anterior margin of the middle segment is irregular in its union with the anterior segment. The second cuneiform bone is smaller, and thus its anterior margin is set back from the other two cuneiform bones. This arrangement creates a mortice into which fits the base of the second metatarsal bones.

The metatarsal bones articulate with the tarsal bones in a unique manner. The second metatarsal bone fits into the mortice between the first

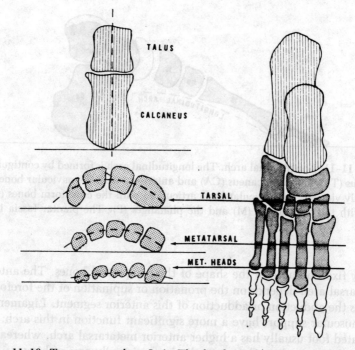

TALUS

CALCANEUS

TARSAL

METATARSAL

MET. HEADS

Figure 11-10. Transverse arches. *Left,* The fixed tarsal and posterior metatarsal arches and the flexible anterior metatarsal arch. *Right,* The receded second cuneiform that creates a mortice for the seating of the base of the second metatarsal.

and third cuneiform bones and thus can move in only one plane: dorsal or plantar flexion. The third, fourth, and fifth metatarsal bases are obliquely shaped to permit rotation upon the adjacent metatarsals. Because of this rotation potential, the forefoot can "cup" and increase the transverse arch. The fixed second metatarsal remains the apex of the arch.

The first metatarsal also rotates about the base of the second metatarsal, but in the opposite direction to complete the arch. The base of the first metatarsal has a cartilaginous surface and is so shaped to permit dorsal and plantar flexion by gliding upon the first cuneiform bone, about which it also rotates.

There are four arches to the foot: the longitudinal arch and three transverse arches. The longitudinal arch is formed by the union of specifically shaped bones that require little ligamentous and muscular support to maintain the arch (Fig. 11–11). The plantar fascia acts as a bowstring that reinforces the arch, as do the intrinsic muscles of the plantar aspect of the foot.

The three transverse arches are the tarsal, posterior metatarsal, and anterior metatarsal. The tarsal and the posterior metatarsal arches are rel-

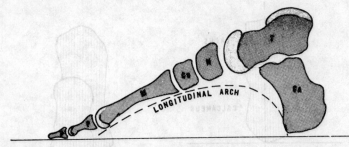

Figure 11-11. Longitudinal arch. The longitudinal arch is formed by contiguity of the talus (T) with the calcaneus (CA) and anteriorly with the navicular bone (N). Medially viewed, the navicular bone articulates with the cuneiform bones (CU), thus with the metatarsals (M) and the phalanges (P). The plantar fascia is not shown.

atively fixed because of the shape of the component bones. The anterior metatarsal arch depends on the pronation or supination of the forefoot as well as the abduction or adduction of this anterior segment. Ligamentous and muscular support have a more significant function in this arch. The supinated foot usually has a higher anterior metatarsal arch, whereas the pronated foot tends to flatten to eliminate virtually any arching.

The forward projection of the metatarsal bones normally follows a sequence of $2 > 3 > 1 > 4 > 5$, with the second metatarsal, which is the longest metatarsal, projecting the furthest.

The phalanges move in the one plane created by flexion and extension (Fig. 11-12). The proximal phalangeal articular surface glides upon the large convex articular surfaces of the metatarsal heads. These metatarsal articular surfaces extend from the plantar surface to the dorsal surface, thus allowing significant dorsiflexion of the toes.

The forward projection of the phalanges differs from that of the metatarsals in that the big toe protrudes forward the farthest and protrusion of each of the other toes declines in sequence (Fig. 11-13).

If the physician has adequate knowledge of the bony anatomy of the foot and ankle, he can examine the foot bimanually and test each joint separately, thus determining joint function and locating the site of pain. The talocalcaneal joint can be tested by fully dorsiflexing the ankle, which immobilizes the talus, and by moving the calcaneus in a varus-valgus direction. The medial and lateral collateral ankle ligaments can be evaluated in a neutral foot position or a position of plantar flexion.

With the hindfoot segment immobilized (heel held firmly in one hand), the subtalar joint can be examined by pronation and supination of the forefoot. Based on the knowledge that the second metatarsal joint moves only in a plane of plantar dorsiflexion and that all the other joints

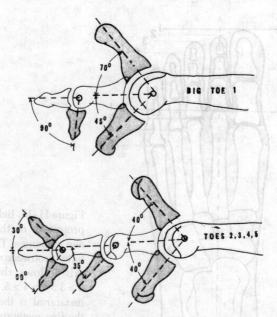

Figure 11-12. Range of motion of phalanges. The phalanges move in only one plane—that created by flexion and extension. *Top*, The big toe (hallux). *Bottom*, The remaining toes have three phalanges.

have rotatory range as well, each metatarsotarsal joint can be tested. The phalanges can be tested individually by fixing the proximal phalanx and moving the more distal joint.

Muscles

The muscles that act upon the foot are the **extrinsic muscles**, which originate from the lower leg and attach to the foot, and the **intrinsic muscles**, which originate and insert within the foot.

The extrinsic muscles include the gastrocnemius soleus group, which allows plantar flexion of the foot and inversion of the forefoot. The lateral or peroneal muscles evert the foot; the medial or posterior tibial muscles allow inversion as well as plantar flexion, and the anterior group or tibial muscles dorsiflex and supinate the forefoot simultaneously.

The muscles act differently in the non-weight-bearing and weight-bearing foot positions. The origin and insertion alternate. In the non-weight-bearing position, the origins are on the tibia and fibula, and insertion and action are upon the foot. Upon weight bearing, however, the origin is on the plantigrade foot, and insertion and action are upon the leg.

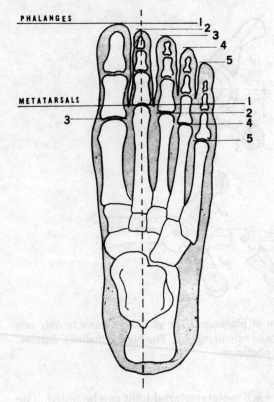

PHALANGES

METATARSALS

Figure 11–13. Relative length of projection of the metatarsals and phalanges. The relative anterior protrusion of the metatarsals follows the pattern of $2>3>1>4>5$. The second metatarsal is the longest, and the first metatarsal is the third longest. The toes have a pattern of $1>2>3>4>5$ in which the first toe protrudes farthest forward followed by each adjacent toe in regular sequence.

The muscles that act upon the toes originate from the tibia, fibula, and interosseous membrane. The long extensor muscle inserts upon the distal phalanx and extends the big toe. The long extensor tendons of the toes divide into a central tendon, which inserts upon the middle phalanx, and two divided slips, which insert upon the distal phalanx. The long extensor muscle extends the toes and assists in everting the foot.

The flexors of the toes are both extrinsic and intrinsic. The long flexor muscle of the great toe originates from the tibia and fibula, and its tendon passes under the medial malleolus to attach upon the distal phalanx of the big toe. This tendon crosses two joints and acts to press the distal phalanx to the floor (Fig. 11–14).

The long flexor muscle of the great toe originates also from the posterior aspect of the tibia. It passes behind the medial malleolus, and attaches to the distal phalanx of all other four toes, but by passing three joints, it acts to grip the floor.

The short portions of all the flexor muscles are intrinsic muscles originating within the plantar aspect of the foot.

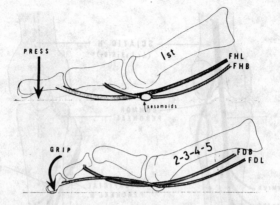

Figure 11–14. Action of the flexor tendons. The tendons of the big toe cross two joints and thus act by pressing the distal phalanx to the floor. The sesamoid bones are incorporated into the tendons of the short flexor muscle of the great toe (flexor hallucis brevis, FHB) and act as a fulcrum for the flexor action. The flexor tendons of the other four toes cross three joints and act to grip the floor. This action is performed by the short flexor muscles (flexor digitorum brevis, FDB) and long flexor muscles of the toes (flexor digitorum longus, FDL).

The nerve supply of the lower leg and foot consist of branches of the sciatic nerve (L_4 to L_5; S_1, S_2, and S_3) and are depicted in Figure 11–15. The tibial nerve proceeds to the intrinsic muscles of the foot by way of the plantar nerve (Fig. 11–16).

EVALUATION OF THE PAINFUL FOOT

The normal foot must conform to the following criteria:

1. It must be pain-free.
2. It must have normal muscle balance.
3. It must have no contractures.
4. The heel must be central.
5. The toes must be straight and mobile.
6. During gait and stance, it must have three sites of weight bearing.

Pain, difficulty in walking, or an awkwardness of gait are the usual problems that cause the patient to seek consultation. Foot pain noted during standing can be considered **static,** and that noted during walking can be considered **kinetic.** The history and evaluation of the foot during relaxation, standing, and walking constitutes an adequate examination. Usually

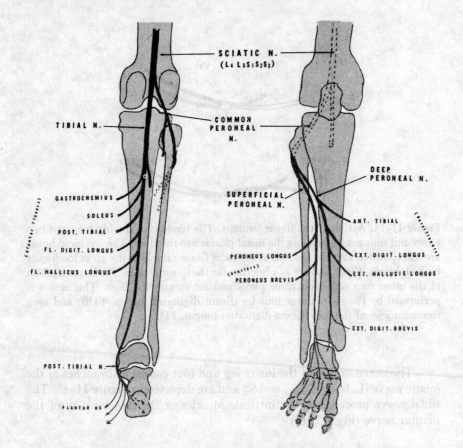

Figure 11-15. Innervation of the leg and foot. The sciatic nerve divides at the popliteal angle to form the tibial nerve and the common peroneal nerve.

the patient can specifically indicate the site of pain, which is discernible visually upon palpation or passive movement.

Since this chapter is directed principally to foot pain, the discussion of neurologic examination through performance of motor function tests has been omitted. Such discussion is included in my book *Foot and Ankle Pain*.[1]

Most painful conditions of the foot originate in the soft tissue: muscles, ligaments, tendons, nerves, blood vessels, and tissues of the joint spaces. In most cases of foot and ankle pain, a local lesion, which may be the result of trauma or stress, is implicated.

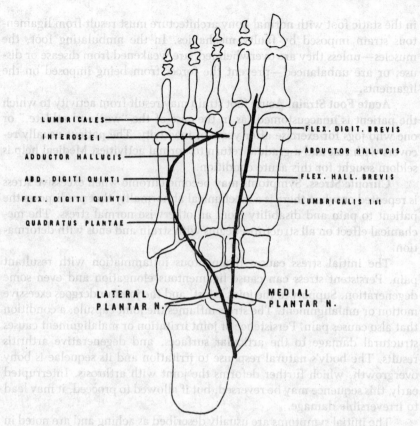

LUMBRICALES

INTEROSSEI

ADDUCTOR HALLUCIS

ABD. DIGITI QUINTI

FLEX. DIGITI QUINTI

QUADRATUS PLANTAE

FLEX. DIGIT. BREVIS

ABDUCTOR HALLUCIS

FLEX. HALL. BREVIS

LUMBRICALIS 1st

LATERAL
PLANTAR N.

MEDIAL
PLANTAR N.

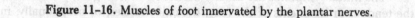

Figure 11-16. Muscles of foot innervated by the plantar nerves.

Foot Strain

Foot strain may be acute, subacute, or chronic. As in so many painful musculoskeletal conditions, the causes may be classified as (1) abnormal stress upon a normal structure, (2) normal stress upon an abnormal structure, or (3) normal stress upon a normal structure that is not at that moment prepared to receive the stress.

The static weight-bearing foot is supported by the configuration of its bony components held together by ligaments. During stance, the muscles of the foot provide no support, neither intrinsic nor extrinsic, even when heavy objects are held by the individual.

During ambulation, however, muscular activity prevents excessive strain upon the supporting ligaments in the moving foot. Therefore, pain in the static foot with normal bony architecture must result from ligamentous strain imposed by faulty mechanics. In the ambulating foot, the muscles—unless they are overwhelmed, are weakened from disease or disuse, or are unbalanced—prevent the forces from being imposed on the ligaments.

Acute Foot Strain. Acute foot strain may result from activity to which the patient is unaccustomed, as in the case of the "weekend athlete," or one who jogs for exercise after years of inactivity. The patient usually recovers with rest and a gradual return to normal activities. Medical help is seldom sought for this acute condition.

Chronic Stress. Symptoms may become chronic when excessive stress is repeated or when there is a mechanical abnormality that predisposes the patient to pain and disability from an otherwise normal stress. The mechanical effect on all structures begins with strain and ends with deformation.

The initial stress causes ligamentous inflammation with resultant pain. Persistent stress can cause ligamentous elongation and even some degeneration. Support to the joint is lost, and the joint undergoes excessive motion or malalignment. The stress inflames the joint capsule, a condition that also causes pain. Persistence of joint irritation or malalignment causes structural damage to the articular surfaces, and degenerative arthritis results. The body's natural response to irritation and its sequelae is bony overgrowth, which further deforms the joint with arthrosis. Interrupted early, this sequence may be reversed, but if allowed to proceed, it may lead to irreversible damage.

The initial symptoms are usually described as aching and are noted in the tendons or ligaments of the foot, the calf muscles, or occasionally, in the anterior musculature of the leg. Deep tenderness of these inflamed tissues is palpable (Fig. 11–17).

The weight-bearing foot is a complex structure with all component parts interdependent (Fig. 11–18). The body weight is transmitted through the tibia upon the talus, which is in turn supported by the calcaneus. The calcaneus is oblique to the horizontal ground surface and thus encourages the forward and medial gliding of the talus. This force further everts the calcaneus and depresses its anterior portion. Because of these changes, the plantar fascia becomes involved in supporting the longitudinal arch and becomes tender.

The increased obliquity of the calcaneus places stress on the medial longitudinal (deltoid) ligament, producing another site of pain. The forward gliding of the talus upon the calcaneus puts stress upon the inferior calcaneonavicular ligament, depressing the navicular bone with further decrease of the longitudinal arch.

As the calcaneus everts (valgus), the forefoot abducts, an action that decreases the two anterior transverse arches. The anterior metatarsal arch,

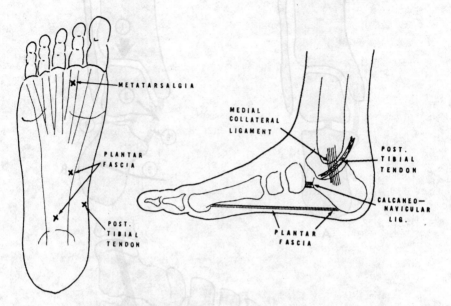

Figure 11-17. Tender areas in foot strain. All the soft tissues that become tender from foot strain are pointed out by patient and palpable by the examiner.

when depressed, splays the forefoot and causes the arch to disappear (Fig. 11-19). Now weight is borne on all metatarsal heads, although weight bearing is not their function, and pain results.

As the calcaneus goes into valgus, the Achilles tendon undergoes adaptive shortening, thus causing further valgus and equinus and further strain on the anterior segments of the foot.

This mechanism and sequence of foot strain can occur in the normal foot that has become deconditioned from prolonged inactivity, bedrest, or immobilization from casting for another, unrelated orthopedic problem. Excessive weight gain, or assuming a new profession requiring prolonged standing after years in a sedentary profession, may be contributing factors. Chronic foot strain, however, is more apt to result in the already pronated foot.

The pronated foot is supported by muscular activity. When this protective muscular action is overwhelmed, the stress is transferred to the ligaments, the joint capsules, and ultimately, the joints themselves. Persistence of the stress converts the reversible functional deformity into a partially or completely irreversible structural deformity. Symptoms vary with the severity or the chronicity of the stress.

In the early phase of strain upon the pronated foot, the muscles acting across the foot attempt to allay the strain upon the ligaments. As the foot

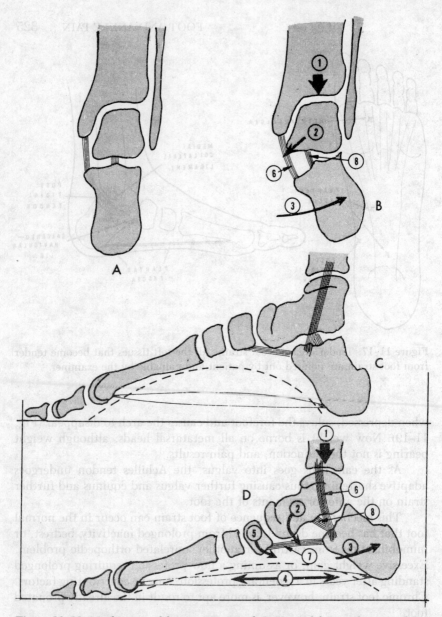

Figure 11–18. Mechanism of foot strain. *A* and *C*, Normal foot with proper bone and joint alignment, a central heel, and good longitudinal arch. *B* and *D*, Stress causes malalignment of structures. Weight-bearing impact of tibia (1) upon talus (2). The talus tends to slide forward and medially upon the calcaneus. Under pressure, the calcaneus everts and rotates posteriorly (3), elongating the longitudinal arch and placing strain upon the plantar fascia (4). The rotating calcaneus depresses the navicular bone (5) by pulling upon the calcaneonavicular ligament (7), which becomes tender. The initial valgus of the heel places strain upon the medial collateral ligament (6) and ultimately upon the talocalcaneal ligament (8).

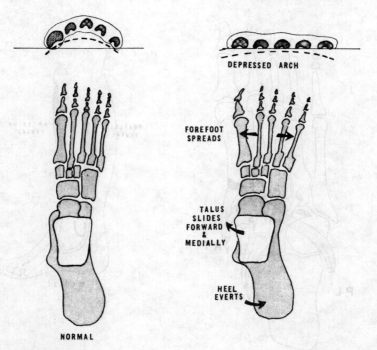

DEPRESSED ARCH

FOREFOOT
SPREADS

TALUS
SLIDES
FORWARD
&
MEDIALLY

HEEL
EVERTS

NORMAL

Figure 11-19. Pronation, which causes splayfoot. As the heel everts, and as the talus slides forward and inward upon the calcaneus, the forefoot abducts and broadens. The anterior metatarsal arch flattens, and the inner three metatarsal heads become weight-bearing.

assumes a more pronated posture, the posterior tibial muscle, an invertor, acts to oppose further pronation. The direction and attachment of the posterior tibialis is depicted in Figure 11-20. Under stress, the posterior tibial tendon becomes tender and can be palpated along its course under and behind the medial malleolus.

The anterior tibial muscle is an invertor as well as a dorsiflexor. During gait, it functions in the latter manner in the swing phase, but has little function in the plantigrade weight-bearing foot.

As the foot goes further into pronation, the lateral evertors shorten to take up the slack. The evertors compose the peroneal muscle group, but with further forefoot pronation, the toe extensors change their alignment and become evertors of the foot as well. In prolonged stress, the evertors may become inflamed and tender.

The talocalcaneal ligament is normally taut in the supinated foot and slack in the pronated foot. As the foot pronates, the tarsal canal deforms. This deformation subjects the talocalcaneal ligament to abnormal stress,

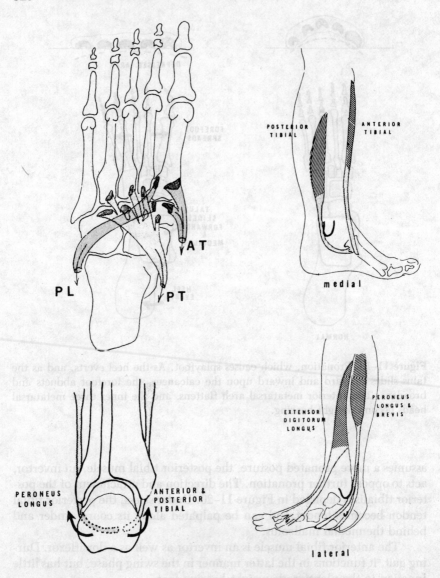

Figure 11-20. Extrinsic musculature of the foot. The origin, direction, and insertion of the extrinsic muscles acting upon the foot are shown. The anterior tibial (AT) and the posterior tibial (PT) muscles are medial muscles that invert the foot. The long peroneal muscle (peroneus longus, PL) everts the foot. The posterior tibial and long peroneal muscles also cause plantar flexion of the foot. The extensor muscles of the toes and the anterior tibial muscle dorsiflex the foot.

and it becomes inflamed. The tenderness here can be palpated by deep pressure into the lateral opening of the canal just anterior to and below the lateral malleolus (see Fig. 11-7).

The plantar fascia becomes elongated as the longitudinal arch flattens. Tender areas may be palpable.

The toe flexors also contribute to maintenance of the longitudinal and transverse arches. Normally, the long extensors extend the distal interphalangeal joints, thus allowing the toe flexors to press the straightened toes against the floor. This action elevates the anterior metatarsal arch simultaneously. In the pronated foot, the everted forefoot causes a malalignment of the toe extensors, which may hyperextend the metatarsophalangeal joint, causing the flexors to claw the toe. The big toe extensor muscle imposes traction on the longitudinal arch (see Fig. 11-39). In this position, the metatarsal heads become more prominent and bear more weight (Fig. 11-21).

Ultimately, the calcaneocuboid and talonavicular joints develop more "play" and sustain capsular and articular irritation with possible pain (Fig. 11-22). Pain originating at these joints can be verified by manually and forcefully everting (pronating) the forefoot while simultaneously immobilizing the heel. Tenderness can be elicited by pressure upon the plantar surface at the calcaneonavicular joint and its ligament.

Treatment. Treatment of foot strain essentially involves treating the pronated foot in the adult. In the acute strain, local and general rest are indicated. Such rest may include avoidance of weight bearing, use of crutches, bedrest with elevation of the foot, and the application of heat or cold. Early in the treatment of strains, ice usually is of value in that it decreases congestion, swelling, and pain. Heat later enhances healing. Many clinicians advocate alternation of heat and cold application.

To rest the joint(s) and ligaments, the foot can be splinted by selective application of adhesive tape. Taping, so valuable in athletic injuries, is too often denied the nonathlete who sustains a severe sprain. Tape with $1/4$ to $1/2$ inch of adhesive-backed sponge rubber can be applied to elevate the portion of the foot desired. In severe sprains, immobilization for several weeks in a plaster cast, with or without a walking heel, is useful.

Local injection of an anesthetic agent with or without steroids is valuable, provided that the exact local anatomic structures are understood. Most ligaments and tendons are readily accessible to injection, and the benefit derived is rewarding. Even the talocalcaneal ligament can be easily injected by entering the canal at the lateral opening, which is palpable directly under and just anterior to the lateral malleolus. Inversion of the forefoot opens the canal. The needle must proceed posteriorly and medially at a 45-degree angle. Injection given in this manner is both safe and valuable.

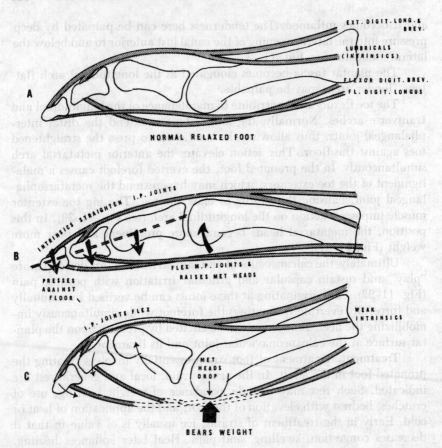

Figure 11–21. Muscular mechanism forming transverse arch. The transverse arch is only a potential arch, not an anatomic arch. The metatarsal heads are raised by the toes being kept straight and the metatarsophalangeal joints being flexed (B) by the long and short flexors. Weakness of the intrinsics permits the toes to bend at the interphalangeal joints; the flexors then increase the flexion at the interphalangeal joints. The metatarsal heads thus bear the full body weight (C).

Injection of joints is also feasible. Most joints of the posterior and middle segments are as accessible to intra-articular injections as are the joints of the anterior segment.

Replacing shoes with proper footwear is valuable. The heel must be low. In pronated feet, a Thomas heel inverts the heel and supinates the foot (Fig. 11–23). The Thomas heel also assists by directing the foot in a slightly toe-in gait. The Thomas heel is preferably modified by elevating the inner border with a ⅛-inch to 3/16-inch wedge. This inner wedge may be extended to include the inner sole of the shoe.

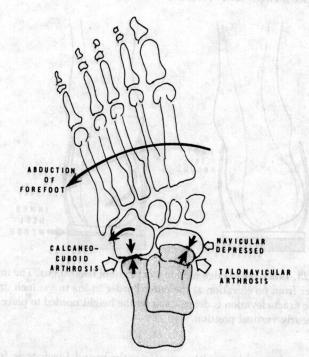

Figure 11-22. The chronically strained foot. The abducted forefoot of the chronically strained foot causes articular changes due to pressure of the abducted cuboid upon the calcaneus. The talus drops and causes pressure upon the superior portion of the navicular bone. Mechanical pressures at joints cause arthrosis.

Other inserts in the shoe, such as those depicted in Figure 11-24, can alter areas of pressure upon the foot or can alter the weight-bearing or ambulatory foot. The metatarsal bar placed across the metatarsal heads avoids pressure upon the heads and decreases the tendency of the toes to hyperflex at the interphalangeal joints and to hyperextend at the tarsometatarsal joint.

Metatarsal pads placed behind the second, third, and fourth metatarsal heads elevate the forefoot and restore the transverse arch. The longitudinal pad insert restores the longitudinal arch and simultaneously inverts the foot to supination.

The shoe must be broad across the forefoot to allow the spread of the anterior segment of the foot. Since most shoe lasts have a toe-in counter they do not conform to the toe-out (abducted) forefoot of the pronated foot. Most shoes also have a tapered last with a pointed toe and thus constrict the forefoot (Fig. 11-25). Custom-made shoes are often necessary when fitting the foot with a standard shoe is difficult.

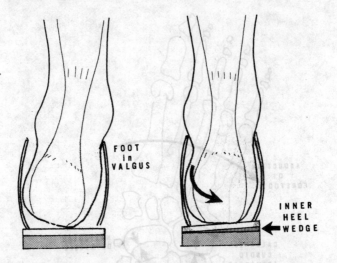

FOOT
in
VALGUS

INNER
HEEL
WEDGE

Figure 11-23. Inner heel wedge used in treatment of heel valgus. The inner wedge should taper from no elevation at the outer border to $1/16$ to $3/16$ inch on the inner border. The exact elevation is determined by the height needed to place the calcaneus in a nearly vertical position.

Often, when a shoe with a sufficiently broad forefoot is found, the counter of the heel is too broad and the customer (i.e., the patient) sacrifices the comfortable forefoot width to acquire a snugger heel. To obviate this problem, the shoe of proper width can be modified by inserting thin felt pads on the lateral and medial aspects of the counter (Fig. 11-26).

The weight-bearing foot is supported mostly by the ligaments and joint capsules supporting the interlocking bones; little support is afforded by the muscles. The muscles, however, do relieve some stress in the walking foot and therefore should be strengthened.

The posterior tibial and the gastrocnemius soleus muscles are foot invertors and are most beneficial in preventing pronation. Standing with feet slightly apart and in a slightly toe-in position, then rising up and down on the toes, strengthens these muscles. Rolling both feet inward to place weight bearing upon the lateral border of the feet, then flexing the toes, also strengthens the invertors.

Practicing gait by walking with a slightly toe-in or exaggerated heel-toe motion is beneficial. A contracted heel cord aggravates pronation; therefore, heel cords should be stretched. The Achilles tendon can be stretched in the following manner. The patient stands a distance from the wall and leans against it. The heel of the weight-bearing foot is held to the floor while the body leans forward. Rising up and down upon the weight-bearing foot while leaning forward gradually stretches the Achilles tendon and strengthens the gastrocnemius muscle (Fig. 11-27).

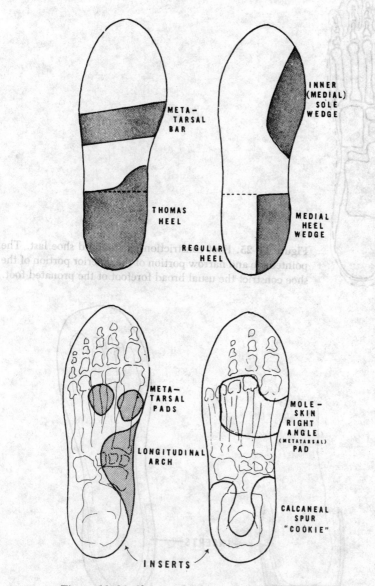

Figure 11-24. Shoe modifications to correct strain factors.

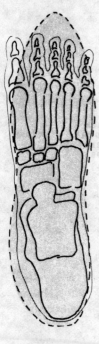

Figure 11–25. Foot constriction by standard shoe last. The pointed toe and narrow portion of the anterior portion of the shoe constrict the usual broad forefoot of the pronated foot.

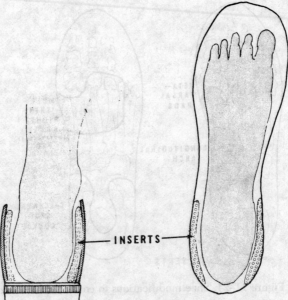

INSERTS

Figure 11–26. Counter insert. When a shoe is sufficiently broad for the splayed forefoot, the counter is too broad to hold the heel. Inserting felt of ⅛- to ¼-inch width inside the shoe makes the heel snug and supported.

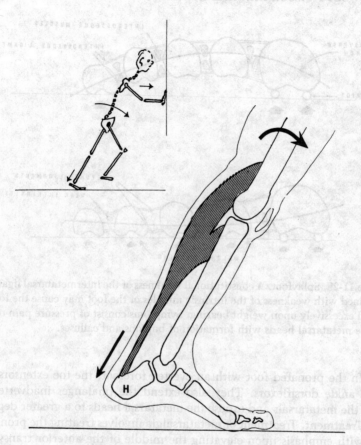

Figure 11-27. Exercise to stretch heel cord.

Metatarsalgia

Metatarsalgia is a condition in which there are pain and tenderness of the plantar heads of the metatarsals. The condition usually occurs when the anterior transverse arch is depressed and causes excessive weight bearing upon the middle (second, third, and fourth) metatarsal heads.

In the normal foot, most weight is borne upon the first metatarsal (big toe) and last metatarsal (fifth toe), which are padded. Five sixths of the weight is calculated as being borne normally by the first metatarsal head.

In the pronated foot, the forefoot spreads and causes the interosseous ligaments to stretch and the transverse arch to depress (Fig. 11-28). Weight is borne upon the second and third heads, which are not adequately padded for this function; therefore, they become tender.

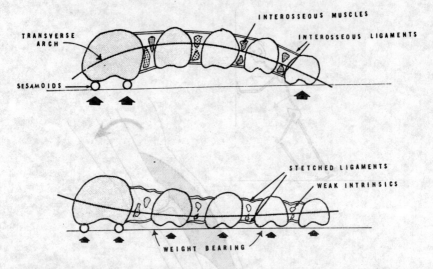

Figure 11-28. Splayfoot. A constitutional weakness of the intermetatarsal ligamen combined with weakness of the intrinsic muscles of the foot may cause the foot to spread excessively upon weight bearing. Symptoms consist of pressure pain of the middle metatarsal heads with formation of bunions and calluses.

In the pronated foot with an everted forefoot, the toe extensors become ankle dorsiflexors. They also extend the phalanges inadvertently upon the metatarsals and expose the metatarsal heads to a greater degree.

Treatment. Treatment of metatarsalgia involves treating the pronated foot with emphasis upon elevating the middle of the anterior transverse arch. This arch is elevated by a pad placed within the shoe under the second and third metatarsal bones, *behind* the metatarsal heads (Fig. 11–29). If the pad is placed under the metatarsal heads, it aggravates the metatarsalgia. The correct placement of the pad is indicated to the patient by pressure of the examiner's thumb at the exact site desired. The patient can then feel where the pad should be and can modify its position within the shoe. Standard pads with adhesive to permit adherence to the shoe can provide an inexpensive support for all the patient's shoes. A Thomas heel with an inner wedge is also indicated.

Morton's Syndrome

A short first metatarsal bone causes excessive weight to be borne by the second metatarsal. This condition is congenital and was first described by Dudley Morton. Morton's syndrome, illustrated in Figure 11–30, consists of (1) an excessively short first metatarsal that is hypermobile in its

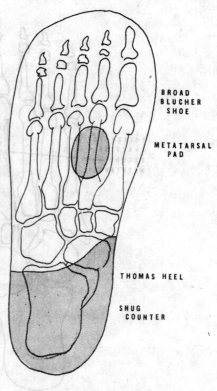

BROAD
BLUCHER
SHOE

METATARSAL
PAD

THOMAS HEEL

SNUG
COUNTER

Figure 11-29. Shoe modifications in treatment of metatarsalgia. Placement of the metatarsal pad to elevate the heads of the second and third metatarsals must be *behind* the metatarsal heads. The shoe with a broad forefoot and a soft upper permits spreading of the foot without cramping, and the soft upper prevents calluses from forming on the dorsum of the toes. If a Thomas heel is used, the counter must be snug to hold the heel centrally.

articulation with the cuneiform bone and with the base of the second metatarsal, (2) posterior displacement of the sesamoids, and (3) thickening of the second metatarsal shaft.

Pain is usually felt at the base of the first two metatarsals and at the *head* of the second. Verification by roentgenography is diagnostic.

Treatment. Treatment of Morton's syndrome requires an orthotic support under the first metatarsal bone to relieve weight bearing upon the second. Other aspects of treating the pronated foot, which have been discussed previously, are also beneficial.

March Fracture

A march fracture, shown in Figure 11-31, is a stress fracture of a metatarsal shaft, the cause of which often involves minimal or no trauma. Its name is derived from the pain noted after a long march. Initial x-ray images may be negative because the fracture is hairline with no displacement of fragments. Later, as callus forms around the fracture, roentgenography reveals that a fracture existed and caused the symptoms. Clinically,

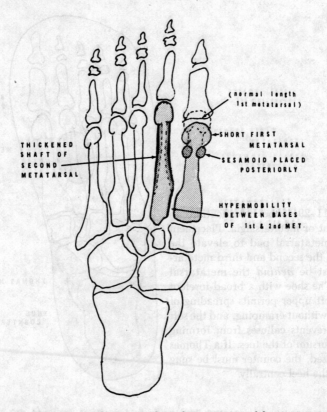

Figure 11-30. Morton's syndrome. A short first metatarsal bone causes excessive weight to be borne by the second metatarsal.

there is tenderness at the middle of the involved metatarsal shaft, and flexion or extension of the toes may be painful.

Treatment. Treatment consists of avoidance of weight bearing or walking, depending upon the severity of the symptoms. When symptoms are severe, a walking cast may be applied for 3 to 4 weeks.

Interdigital Neuritis

Morton's Neuroma. The most common type of interdigital neuritis is Morton's neuroma (Fig. 11-32). This neuroma consists of a painful fusiform swelling of a digital nerve. The most common site is the nerve between the third and fourth toes, but it can also occur between the second and third metatarsals. The neuroma is usually located where the interdigital nerve separates (between the metatarsal heads) into branches to the two contiguous digits.

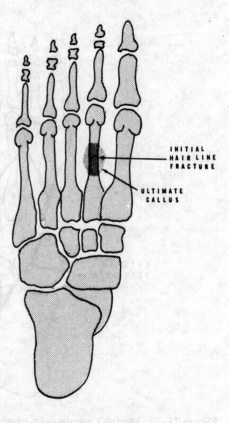

INITIAL
HAIR LINE
FRACTURE

ULTIMATE
CALLUS

Figure 11-31. March fracture of the second metatarsal. Often, the initial fracture is not observed in routine roentgenography, or it appears as a hairline fracture. Within three weeks, after persistent pain, swelling, and tenderness, a callous formation indicative of healing becomes evident radiologically. This may be the first diagnostic sign. Displacement of fragments is rare.

This condition is found most often in middle-aged women. A characteristic history usually consists of the patient's desire to take off the shoes and massage the metatarsal area. Getting off one's feet does not afford relief as does removing the shoes.

On examination, pain is reproduced by pressure *between* the metatarsal heads (whereas in metatarsalgia, tenderness is found by pressure on the plantar surface of the metatarsal heads). Numbness or hypalgesia may occur in the contiguous areas of the toes innervated by that interdigital nerve. Compressing the metatarsal heads together may elicit pain.

For treatment of the condition, mere prescription of adequately broad shoes to permit spread of the forefoot may suffice. A metatarsal pad placed behind the heads to elevate the transverse arch provides some benefit. An injection of an anesthetic agent with steroid that provides relief is diagnostically and frequently therapeutically valuable. The metatarsal heads are easily palpated, and the injection is given from the dorsum of the foot directly into the interdigital area. Persistence of symptoms warrants surgical excision.

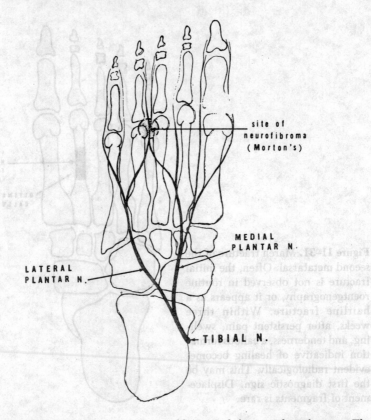

Figure 11-32. Morton's neuroma, a neurofibroma of the interdigital nerve. The most frequent site is the third branch of the medial plantar nerve as it merges with the lateral plantar nerve to form the digital nerve between the third and fourth toes. Pain occurs in the area, and hypalgesia may occur in the opposing areas of the foot.

Neuritis at the Transverse Metatarsal Ligament. Another type of interdigital neuritis may occur from entrapment of the interdigital nerves as they pass the transverse metatarsal ligament (Fig. 11-33). This particular type of neuritis is caused by hyperextension of the toes. Clinically, pain and tenderness are induced by pressure *between* the metatarsal heads and by hyperextension of the toes. Treatment consists essentially of evaluating the patient's activities and advising the patient to avoid any position in which the toes hyperextend.

Posterior Tibial Neuritis. The posterior tibial nerve may be entrapped in its passage from behind and under the medial malleolus to a point under the laciniate ligament (Fig. 11-34). This area, which is composed of a bony depression that is situated behind the medial malleolus and that is

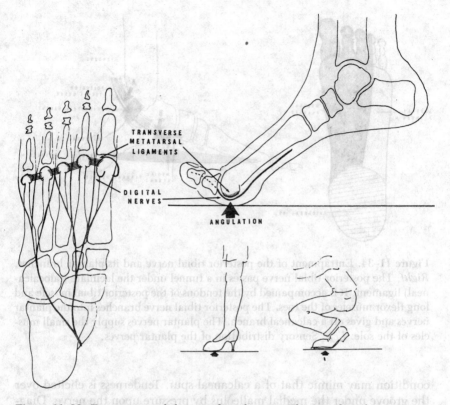

Figure 11–33. Entrapment of the interdigital nerve. The interdigital nerves provide sensation to the toes. They originate from the plantar nerves in the sole of the foot and pass across the transverse metatarsal ligament to the dorsum of the toes. If they are angulated at the ligament by posture and weight bearing, pain in the foot and numbness of the toes may result.

covered by the laciniate ligament, forms the tarsal tunnel. In addition to the posterior tibial nerve, the tendons of the posterior tibial muscle, the long flexor muscles of the toes, and the tibial artery and veins are contained in this region. The posterior tibial nerve carries sensation to the sole of the foot and innervates the intrinsic muscles of the foot.

The nerve may be injured, compressed, or entrapped from a direct fall upon one's feet, from standing on a ladder rung with soft shoes or bare feet, from acute pronation of a normal foot, or from accentuation of an already pronated foot. Ill-fitting longitudinal arch supports may also cause the pressure.

Diagnosis is suggested by a history of "burning" in the plantar distribution of the nerve. This burning can be felt during standing or sitting and is *not* related to weight bearing. If the calcaneal branch is affected, the

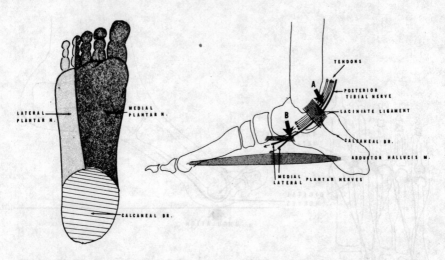

Figure 11–34. Entrapment of the posterior tibial nerve and its plantar branches. *Right*, The posterior tibial nerve passes in a tunnel under the laciniate (talocalcaneal) ligament. It is accompanied by the tendons of the posterior tibial muscle and long flexor muscles of the toes. The posterior tibial nerve branches into the plantar nerves and gives off a calcaneal branch. The plantar nerves supply the small muscles of the sole. *Left*, Sensory distribution of the plantar nerves.

condition may mimic that of a calcaneal spur. Tenderness is elicited over the groove under the medial malleolus by pressure upon the nerve. Diagnosis can be verified by injecting an anesthetic agent into the tunnel, which should relieve the pain. Electromyographic conduction times are specifically diagnostic if prolonged. Sensory testing by touch or pin scratch may be used to outline the dermatomal hypesthesia. In addition, tapping of the tunnel may produce Tinel's sign if the proximal flexors are weak.

Treatment by mere correction of the foot pronation through taping, casting, shoe correction, and gait modification may be successful. When tenosynovitis of the tunnel exists in conditions such as rheumatoid arthritis, and the inflamed tendons compress the nerve, oral steroids or local injection of steroids are beneficial.

Anterior Tibial Neuritis. Anterior tibial neuritis may result from injury to the distal branch of the deep peroneal nerve (Fig. 11–35). This nerve branch accompanies the dorsal artery and becomes superficial below the cruciate crural ligament. The sensory distribution of the nerve is the dorsal cleft between the first and second toes.

The common injury is caused by direct pressure on the nerve, usually resulting from an ill-fitting shoe. Local injection of a Novocain derivative may be diagnostically and therapeutically beneficial. Correction of the pressure prevents recurrence.

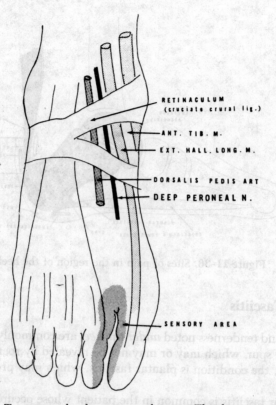

RETINACULUM
(cruciate crural lig.)

ANT. TIB. M.

EXT. HALL. LONG. M.

DORSALIS PEDIS ART

DEEP PERONEAL N.

SENSORY AREA

Figure 11–35. Trauma to the deep peroneal nerve. The deep peroneal nerve becomes superficial as it emerges below the cruciate crural ligament. There it is vulnerable to trauma causing pain and numbness in the area shown in the diagram.

Referred Pain

Pain can be referred to the foot-ankle region from abnormal conditions in spinal nerve roots, such as a herniated disk or a spinal cord tumor. This referral pattern has been thoroughly discussed in Chapter 3, but it is mentioned here to alert the physician to that possibility when there is pain in the foot-ankle region.

The Painful Heel

Pain in the region of the heel may (1) arise from the tissues behind and under the calcaneus, (2) arise within the bones and joints of the heel, or (3) be referred to the heel region from a distant site (Fig. 11–36).

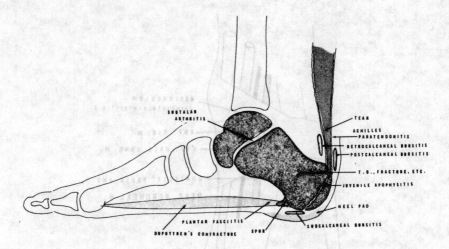

Figure 11-36. Sites of pain in the region of the heel.

Plantar Fasciitis

Pain and tenderness noted *under the heel* are commonly attributed to a calcaneal spur, which may or may not be revealed by roentgenography. Frequently, the condition is plantar fasciitis, which may precede a calcaneal spur.

Plantar fasciitis is common in the patient whose occupation requires long periods of standing or walking, especially when the patient is unaccustomed to such activity. The condition is more prevalent in people with pronated feet, a condition that places stress upon the longitudinal arch. Plantar fasciitis appears to be increasing with the advent of jogging for cardiopulmonary exercise, because joggers may not be sufficiently conditioned to jog in soft shoes and on hard surfaces. Males are more prone to acquiring plantar fasciitis.

This condition may be considered a tendofascioperiosteal irritation. The plantar fascia attaches by tendinous insertion to the periosteum of the calcaneus. The pathologic process probably involves a minor tear or stretching of the plantar tendon fibers with avulsion of the periosteum from the bone. Subperiosteal inflammation occurs. In time, the damage is repaired by fibrous tissue, and then calcium deposit, ultimately forming a spur (Fig. 11-37). The initial pain and discomfort are probably due to inflammation of the fascial, tendinous, and periosteal soft tissues, and only in later stages is the pain due to the spur.

The presenting complaint is pain and tenderness beneath the anterior portion of the calcaneus, with the pain radiating into the sole or into the calcaneal pad. The examination reveals deep tenderness of the anterome-

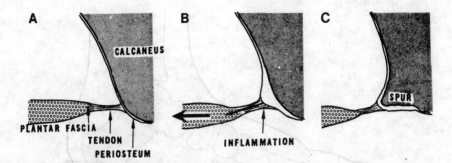

Figure 11–37. Mechanism of plantar fasciitis (heel spur). *A,* Normal relationship and attachment of the plantar fascia to the calcaneus. *B,* Traction upon the fascial tendinous portion to the periosteum separates the periosteum from the heel, and resultant inflammation causes pain. *C,* Subperiosteal invasion by inflammatory tissue and ultimate calcification into a spur. This process may be asymptomatic.

dial aspect of the calcaneus, which is the site of attachment of the plantar fascia. Initially, x-ray images do not reveal any pathologic process. In chronic recurrent episodes of acute fasciitis, the results of x-ray studies may remain negative, and spurs may be noted in patients who are asymptomatic. Therefore, the diagnosis is clinical.

Treatment. Treatment is primarily aimed at relieving local pain, and secondarily aimed at relieving tension upon the plantar fascia or completing the tear, as pain originates at this site. Many patients with plantar fasciitis have tight heel cords and tight gastrocnemius and soleus muscles. Exercises for stretching the heel cord and orthotic shoe inserts to correct the pronation are mandatory.

Injection of an anesthetic agent containing a soluble steroid into the painful area is diagnostically as well as therapeutically useful. The injection may be administered directly into the area through the heel pad or by an approach from the lateral or medial aspect of the heel pad (Fig. 11–38). When the injection is given directly through the heel pad, the point of the needle is placed upon the skin, held momentarily, and then inserted until it touches the bone. Once upon the bone it is withdrawn 1 mm and slowly angled anteriorly until the anterior margin of the calcaneus is reached. This margin is the point of attachment of the plantar fascia upon the periosteum of the anterior margin of the calcaneus. It is here that the plantar fascia imposes traction upon the periosteum, which causes the pain.

Raising the heel of the shoe ¼ to ½ inch decreases the tension upon the calcaneus by removing tension upon the Achilles tendon from the equinus foot position. Placing a sponge rubber insert in the shoe may be beneficial.

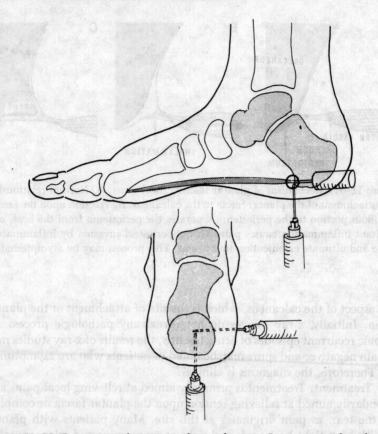

Figure 11–38. Injection technique in plantar fasciitis. Injection can be administered directly into the point of maximum tenderness of the plantar fascia through the heel pad. The fascia inserts into the calcaneus in this area. The site can be reached from the lateral or medial approach, but such localization is less accurate.

Forceful stretching of the plantar fascia after the injection by active dorsiflexion of the foot with the toes hyperextended has been beneficial in my experience (Fig. 11–39).

Surgical removal of the spur, with stripping of the plantar fascia, has been advocated, but surprisingly, there have been many recurrences after this type of surgery.

Painful Heel Pad

Pain and tenderness may occur in the calcaneal pad. This pad is composed of fatty and fibroelastic tissue formed by fibrous septa, or compart-

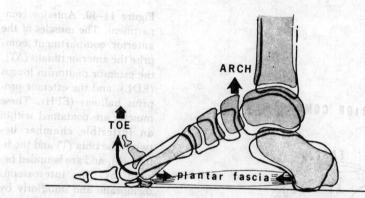

Figure 11–39. Effect of the toes upon the longitudinal arch. Full extension of the big toe exerts traction upon the plantar fascia, which causes elevation of the longitudinal arch.

ments. These compartments act as compressible shock absorbers that bear weight during standing and walking.

The elasticity of the pad decreases with age, and gradually, the calcaneus becomes weight-bearing without the protection of the pad. There are persons who are born with inadequate pads and suffer pain for most of their active lives.

Examination reveals the inflamed or inadequate pads and tenderness over the exposed calcaneus. In chronic conditions, the bone may develop exostosis, which aggravates the painful situation.

Treatment varies with the acuteness or chronicity of the condition. In acute irritation and inflammation of the heel pad, infiltration of a Novocain derivative under the pad may relieve the symptoms. A sponge-rubber heel pad may be inserted into the shoe as a substitute for the calcaneal pad. Elevation of the shoe heel may beneficially transfer weight anteriorly to the calcaneus. Weight reduction and changing the gait pattern have therapeutic value.

Anterior Compartment Syndrome

The anterior compartment is an area beneath the tight fascia that aneriorly invests the muscles between the tibia and fibula (Fig. 11-40). These muscles are the anterior tibialis, extensor proprius longus, and extensor digitorum longus.

Anterior compartment syndrome is usually caused by excessive physical exertion from running, jogging, or tennis performed repeatedly and for long periods of time. The condition also occurs often in poorly conditioned

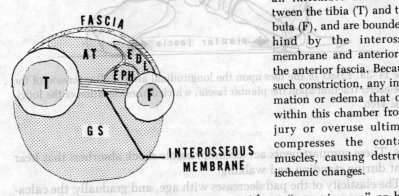

ANTERIOR COMPARTMENT

FASCIA

AT   EDL
EPH

T   F

GS

INTEROSSEOUS
MEMBRANE

Figure 11–40. Anterior compartment. The muscles of the anterior compartment comprise the anterior tibialis (AT), the extensor digitorum longus (EDL), and the extensor proprius hallucis (EPH). These muscles are contained within an inflexible chamber between the tibia (T) and the fibula (F), and are bounded behind by the interosseous membrane and anteriorly by the anterior fascia. Because of such constriction, any inflammation or edema that occurs within this chamber from injury or overuse ultimately compresses the contained muscles, causing destructive ischemic changes.

people who undergo stressful exercise without "warming up" or before attaining sufficient flexibility.

This condition is also known as "shin splints." Pain and tenderness are felt in the region of the anterior lower leg, and any attempt at dorsiflexion of the foot and ankle causes pain.

Edema occurs within the compartment. The inflexible overlying fascia prevents expansion, and the fluid accumulation within the compartment causes muscular ischemia. The pressure within the compartment has been measured and found to be many times the normal pressure.

The usual acute incidence of pain and tenderness subsides within a few days provided that the activity is stopped, the leg is elevated, ice packs are applied to the front of the leg for 15 minutes hourly, a pressure dressing is applied, and oral anti-inflammatory medication is administered. It is important to institute care early as prolongation of the irritating activities combined with increased edema within the compartment can lead to ischemic necrosis of the muscles, and ultimately, paresis of the dorsiflexors of the foot and ankle.

Some clinicians advocate injecting steroids into the compartment *immediately under the fascia*, not into the muscle bundles, when signs and symptoms are chronic and early care has not produced the desired response. If much tension is present within the compartment, a fasciotomy should be considered early for full decompression of the muscles of the compartment. Once ischemic necrosis of the compartment occurs, muscle recovery is extremely limited.

Achilles Tendinitis

The Achilles tendon may become tender. Because the Achilles tendon does not have a synovial sheath, this condition cannot be considered to be tenosynovitis. Inflammation, usually resulting from trauma or stress, occurs in the loose connective tissue about the tendon known as the paratenon.

The tendon is found to be tender when squeezed by the examiner, and it may appear to be thickened or swollen. Pain is aggravated by acute stretching from forceful dorsiflexion of the ankle. Running, jumping, and dancing aggravate the symptoms.

Retrocalcaneal bursitis may be present and may cause similar symptoms. Inflammation of the bursa existing between the Achilles tendon and the skin usually results from ill-fitting shoes in which the shoe counter has caused pressure and irritation.

Treatment of Achilles tendinitis consists of 4 weeks of immobilization of the foot and ankle using a short-leg walking cast. When distension in the bursa is visible, and when fluid is present, aspiration followed by injection of steroids is beneficial. Correction of the shoe irritation is indicated; such correction may require cutting out the offending portion of the shoe. Moleskin tape may be placed on the Achilles tendon, at the site of chronic irritation to prevent further irritation.

Rupture of the Achilles Tendon

Rupture of the Achilles tendon may occur from stress or injury. It may occur from (1) direct trauma to the tendon such as a direct kick sustained during an athletic activity, (2) abrupt stretching of an already fully stretched Achilles tendon, or (3) forceful ankle dorsiflexion when the ankle is relaxed and the patient is unprepared for the stress. Tears of the Achilles tendon occur most often in men between the ages of 40 and 50, especially when strenuous activities are undertaken after years of sedentary living.

Tearing usually occurs in the narrowest portion of the tendon, approximately 2 inches above its point of attachment (Fig. 11–41). The tear may be partial or complete. Most partial tears ultimately become complete.

The person who sustains an Achilles tear experiences acute agonizing pain in the lower calf region, which immediately renders walking impossible. The patient cannot rise onto the toes if the tear is complete.

Examination is best performed with the patient kneeling on the examining table, with the feet hanging over the edge. A gap in the tendon can often be palpated, and the gastrocnemius and soleus muscles are retracted. Ecchymosis may be found around the heel. If the tear is complete, the

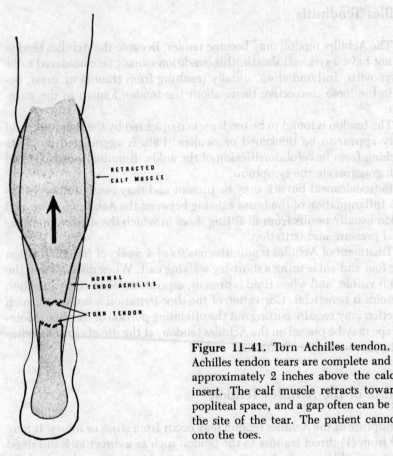

RETRACTED
CALF MUSCLE

NORMAL
TENDO ACHILLIS

TORN TENDON

Figure 11-41. Torn Achilles tendon. Most Achilles tendon tears are complete and occur approximately 2 inches above the calcaneal insert. The calf muscle retracts toward the popliteal space, and a gap often can be felt at the site of the tear. The patient cannot rise onto the toes.

injured ankle can be dorsiflexed to a higher degree than can the normal ankle. Squeezing the normal Achilles tendon causes reflex ankle plantar flexion, but in a complete tear, this reflex action does not occur. The test for this reaction has been termed Simmond's test.

The Thompson test[2] is similar. With the patient prone, the hips and knees extended, and the feet hanging freely over the end of the examining table, a medial and lateral pressure is applied to the belly of the gastrocnemius muscle. Diminished plantar flexion, as compared with plantar flexion of the normal side, indicates possible rupture of the Achilles tendon or of the gastrocnemius and soleus muscles.

Partial tears, which are painful initially, stop causing pain when they are completed. Inability to walk or to rise onto the toes may be the only residual effect. In immediate complete tears, the pain may be momentary, and the patient may merely experience a sudden fall to the ground as the calf muscle gives out.

Treatment. Treatment of the torn Achilles tendon has varied in recent years. Surgical repair followed by casting has had varying success and is still considered by many orthopedists to be the preferable approach in the young person who intends to pursue an athletic career.

A nonsurgical approach of casting the foot for 8 weeks in the position of plantar flexion had led to satisfactory healing of the Achilles tendon with good functional result. The *forced* equinus position must be avoided, and no less than 8 weeks of casting are required for effective treatment. After the cast is removed, a 2.5-cm heel lift should be worn, and the patient should be cautioned about the possibility of falling. Use of crutches for a few days is beneficial.

Active gastrocnemius-strengthening exercises are instituted. The healed tendon remains thicker and the patient cannot rise as high onto the toes as is possible on the normal side, but regular activities can be resumed. This form of treatment is based on the Achilles tendon's ability to regenerate itself and reunite.

Plantaris Tear

A condition characterized by acute pain in the calf experienced during physical activity that may mimic that of a tendon tear is the tearing of the plantar muscle or its tendon.

The plantar muscle originates from the posterior area of the femur just above the lateral epicondyle. The muscle is only 4 to 6 cm in length and becomes a long tendon that descends the entire leg to attach to the calcaneus. It courses between the gastrocnemius and soleus muscles.

The tear occurs during strenuous activity, and pain may be agonizing though brief. Ecchymosis may result in the calf area or migrate to the Achilles area between the tendon and the tibia. The patient can rise onto the toes although pain may occur at first and suggest a tear of the gastrocnemius or soleus muscle. Passive ankle dorsiflexion is not increased as in a tear, but may be restricted because of spasm. Deep tenderness of the calf may occur.

Treatment consists of wrapping the calf with an ace bandage and restricting activities as dictated by pain. Residual disability does not occur.

Fractures of the calcaneus must always be suspected in a painful severe injury and can be verified by roentgenography. Detailed discussion of calcaneal fractures are beyond the scope of this book.

PAINFUL ABNORMALITIES OF THE TOES

Hallux Valgus

Hallux valgus, frequently grouped with the painful condition termed bunions, is the most common painful deformity of the big toe. In essence, hallux valgus is lateral deviation of the proximal phalanx upon the first metatarsal (Figs. 11-42 and 11-43).

Hallux valgus is a complex condition comprising (1) lateral angulation of the first toe toward the second toe, (2) enlargement of the medial portion of the first metatarsal head, and (3) inflammation of the bursa over the medial aspect of the metatarsophalangeal joint.

The symptoms of this condition affect mostly older women who have had pronated feet with broadened forefoot and depressed metatarsal arch. Bunions are attributed in part to wearing shoes with narrow pointed toes,

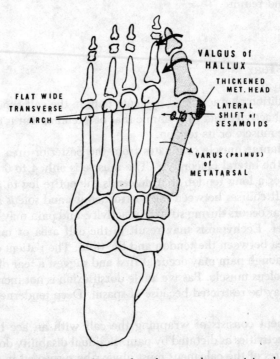

Figure 11-42. Major bony and articular changes characterizing hallux valgus. Hallux valgus is essentially a subluxation of the two phalanges of the big toe in a valgus direction. The first metatarsal deviates in a varus direction, and the sesamoids are thus shifted laterally.

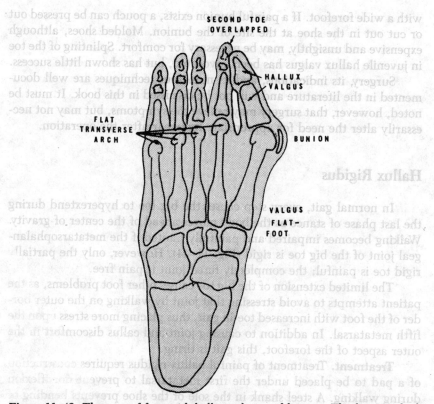

Figure 11–43. The type of foot with hallux valgus and bunion. The foot in which hallux valgus predominates is a broad forefoot with a depressed transverse metatarsal arch and a flattened longitudinal arch. The valgus big toe overrides or underlays the second toe which may be secondarily a hammer toe. A swollen inflamed bursa may overlie the enlarged head of the first metatarsal.

which constrict the forefoot, and high heels, which ensure the foot's being forced into the shoe.

The basic hallux valgus is considered to be a congenital condition. Metatarsus primus varus (medial deviation of the first metatarsal) in childhood may predispose the foot to later hallux valgus. Abnormality of the convexity of the first metatarsal head and muscle imbalance have also been implicated.

Treatment. Treatment of hallux valgus must be individualized and depends on the age of the patient, the severity and duration of symptoms, and the degree of deformity. Many patients with severe deformity are asymptomatic except for the difficulty of finding comfortable or attractive footwear.

Nonsurgical treatment consists of utilizing all the treatment components advocated for the pronated foot, with attention especially to shoes with a wide forefoot. If a painful bunion exists, a pouch can be pressed out or cut out in the shoe at the site of the bunion. Molded shoes, although expensive and unsightly, may be necessary for comfort. Splinting of the toe in juvenile hallux valgus has been advocated, but has shown little success.

Surgery, its indications, and its numerous techniques are well documented in the literature and will not be discussed in this book. It must be noted, however, that surgery may relieve the symptoms, but may not necessarily alter the need for appropriate footwear after the operation.

Hallux Rigidus

In normal gait, every step causes the big toe to hyperextend during the last phase of stance as the body moves ahead of the center of gravity. Walking becomes impaired and painfully limited if the metatarsophalangeal joint of the big toe is rigid (Fig. 11–44). However, only the partially rigid toe is painful; the completely fused joint is pain free.

The limited extension of the big toe causes other foot problems, as the patient attempts to avoid stressing that joint by walking on the outer border of the foot with increased toe-in gait, thus placing more stress upon the fifth metatarsal. In addition to causing joint and callus discomfort in the outer aspect of the forefoot, this gait is tiring.

Treatment. Treatment of painful hallux rigidus requires construction of a pad to be placed under the first metatarsal to prevent dorsiflexion during walking. A steel shank in the sole of the shoe prevents bending of the shoe last, and a rocker sole added to the sole permits the foot to roll over the rocker during gait without extending the big toe (see Fig. 11–44).

Surgical intervention ranges from resection of the joint and remodeling the metatarsal head to replacement of the joint with an orthosis.

Hammer Toe

The hammer toe is a fixed flexion deformity of the interphalangeal joint with hyperextension of the metatarsophalangeal joint (Fig. 11–45). This condition may be congenital or acquired. A painful callus may form on the dorsum of the middle interphalangeal joints or on the tip of the distal phalanx.

Treatment. Treatment may be required to eliminate pain or even ulceration at the callus sites, or may be desired to permit wearing of normal foot gear. To avoid pressure over the protruding joints, a bulge is created in the shoe at the site, the offending portion of the shoe is cut out, or molded shoes are prescribed.

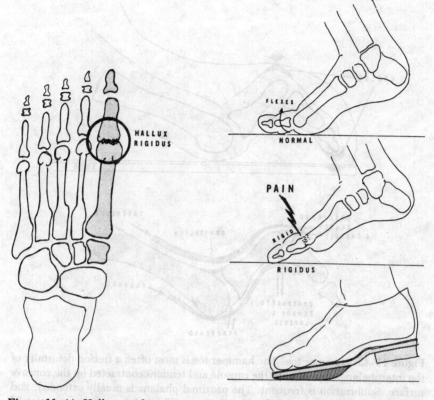

Figure 11-44. Hallux rigidus. As a result of damage of the metatarsophalangeal joint, which becomes rigid, the toe cannot flex on toe-off of the gait, and pain can occur at each step. *Bottom,* Treatment consists of preventing stress on the rigid toe by placement of a steel plate in the shoe sole to prevent bending and a rocker bar to permit painfree gait.

Physical therapy involving manual stretching of the toes and exercise of the imbalanced muscle is usually futile. Surgical intervention is effective and should be sought in severe or disabling instances of hammer toe.

ANKLE INJURIES

Sprain is the most common painful injury to the ankle. The injury may vary from a simple strain, involving mere elongation of the ligaments and microtrauma, to tearing of the ligamentous fibers with or without avulsion of the bones to which they attach. The most severe injury is frac-

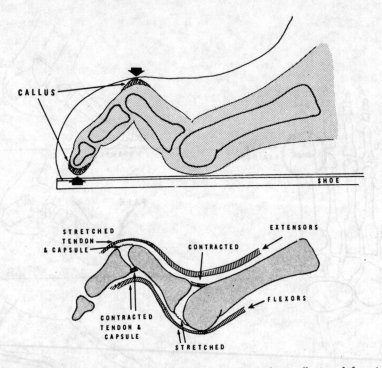

Figure 11–45. Hammer toe. The hammer toe is most often a flexion deformity of the interphalangeal joint with the capsule and tendons contracted on the concave surface. Subluxation is frequent. The proximal phalanx is usually extended, and the distal phalanx flexed and flexible. Pressure and friction result in formation of painful calluses.

ture dislocation of the ankle. The severity of a sprain is so frequently unrecognized that the statement of Watson-Jones must be heeded: "It is worse to sprain an ankle than to break it." This statement implies that fracture receives adequate treatment while sprain is neglected or given inappropriate care.

The lateral and medial collateral ligaments stabilize the ankle while permitting dorsiflexion and plantar flexion. The talus is firmly seated within the ankle mortice in full dorsiflexion, but as it presents its narrower width in plantar flexion, it becomes mobile in a mediolateral direction. The neutral position of the ankle, and especially its position in plantar flexion, expose the ligaments to the possibility of strain and sprain.

All ligaments of the ankle are supplied with sensory nerves and thus are potential sites of pain. When ligaments are stretched, a reflex muscle spasm can occur. Sprain is defined as "a joint injury in which some of the fibers of a supporting ligament are ruptured but the continuity of the ligament remains intact."[3]

Most ankle sprains occur when the ankle is in an unstable position and the stress is borne by the ligaments. The ankle is most unstable when the foot is in plantar flexion and both medial or lateral collateral ligaments are exposed to stress. With the foot plantigrade and fixed to the ground, the superincumbent body weight combined with leg rotation about the ankle can result in a sprain. This maneuver is exemplified by the football player whose cleated foot becomes fixed in the ground as he runs, causing him to turn his ankle and sustain a severe sprain.

The most common ligamentous injury is caused by inversion stress on the lateral ligaments. If the foot is in plantar flexion at the time of the stress, the anterior talofibular ligament is affected, whereas if the foot is in a neutral position, the calcaneofibular ligament sustains the injury.

Various degrees of sprain may occur. The ligament may be overstretched (strained) without disruption of the integrity of the fibers. Usually, this injury is minor, and recovery is rapid and complete. Alternatively, fibers may be torn, which constitutes a sprain; this injury is more severe and requires a longer healing period. Fiber tears may be partial or complete. In severe injuries, the ligament may essentially remain intact but avulse a small fragment of the bone to which it was attached.

Ankle sprains have been classified according to the following:

Grade I: Partial interstitial tearing of ligament.
Grade II: More severe but incomplete tearing, with gross stability retained.
Grade III: Gross instability of the ankle.

Another classification has been proposed:

Grade I: Involves only the lateral ligaments.
Grade II: Involves both medial and lateral ligaments.
Grade III: Involves both lateral and medial ligaments *and* the distal tibiofibular pseudoarticular (interosseous) ligament.

The simple strain does not impair joint stability whereas a sprain with fiber disruption is in essence a self-reduced subluxation with residual instability. All sprains must be assumed to involve ligamentous damage and should be treated according to this assumption.

Eversion injuries, which impose stress on the medial collateral ligament, usually cause bony damage rather than a sprain or tear of the medial collateral ligament, which occurs in inversion injuries. This difference is due to the strength of the medial ligament.

Diagnosis is most accurate when the patient or a witness to the injury has a clear recollection of the details of the incident. Hours or days after the incident, memory of details is vague, and the swelling and ecchymosis are less specific and localized.

During the examination, tenderness over the involved ligament can be elicited. In a mild sprain, the injured foot does not invert to a greater degree than the normal foot, and no gap can be palpated between the foot and the malleolus. Protective spasm of the muscles may have occurred, however, preventing excessive inversion of the foot and thus causing the examiner to underestimate the severity of the injury.

Roentgenography should be performed for every significant or questionable ankle sprain. Studies comparing forced inversion of the involved ankle with forced inversion of the normal ankle are preferable. An abnormal tilting of the talus is revealed in images of a severely sprained ankle (Fig. 11–46).

Novocain can be injected into the tender area to overcome spasm before performing roentgenography, but usually gentle manual inversion accomplishes the same result. Any degree of talar tilting as compared with the normal contralateral ankle suggests ligamentous tear.

Severe injuries, especially those of the medial ligaments, can separate the ankle mortice and tear the tibiofibular ligament (Fig. 11–47). This type of tear widens the ankle mortice and leads to marked instability. Degenerative changes can result in later life from such an injury.

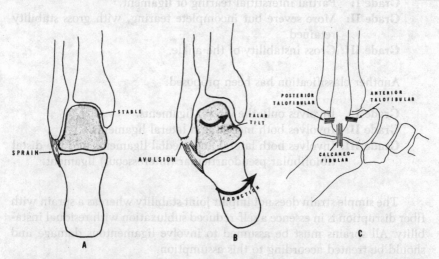

Figure 11–46. Lateral ligamentous sprain and avulsion. A, Simple sprain in which the ligaments remain intact and the talus remains stable within the mortice. B, Avulsion of the lateral ligaments causing the talus to become unstable and tilt within the mortice when the calcaneus is adducted. C, Lateral ligaments of the ankle. The anterior talofibular and calcaneofibular ligaments are the ligaments most frequently involved in inversion injuries.

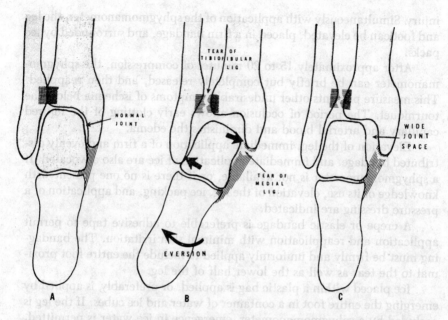

Figure 11–47. Avulsion of medial ligament and anterior tibiofibular ligaments from eversion injury to the ankle. *A*, Normal ankle mortice with the talus fitting snugly between the malleoli. *B*, Eversion stress separates the malleoli (*arrows*) and tears the anterior tibiofibular ligaments and the medial deltoid ligament. Once the ankle has returned to its neutral position (*C*), a wide space remains between the talus and the medial malleolus, which is a diagnostic sign revealed by roentgenography. The ankle remains unstable with the wide mortice.

Treatment

The immediate care of the sprained ankle requires prevention of swelling. The initial treatment usually advocated involves elevation of the limb and application of ice. Swelling occurs because of vascular injury with effusion and either microscopic or macroscopic hemorrhage. Occlusion of the arterial supply to the injured part appears to be physiologic. If swelling occurs, effusion and hemorrhage distend the joint, stretching the ligaments further and predisposing the patient to later formation of adhesions.

Curtailing blood supply to the ankle is a feasible element of treatment. Therefore, immediately upon an injury being sustained and observed, a sphygmomanometer should be applied to the leg immediately below the knee and compressed to pressure exceeding the arterial pressure. This procedure occludes capillary circulation and prevents hemorrhage and edema. It also permits careful and more leisurely examination of the injury. Simultaneously with application of the sphygmomanometer, the leg and foot can be elevated, placed in a firm bandage, and surrounded by ice packs.

After approximately 15 to 20 minutes of compression, the sphygmomanometer can be briefly but completely released, and then reapplied. This measure prevents other undesirable symptoms of ischemia below the tourniquet. The period of occlusion allows early clotting of the injured capillary and arterial blood and diminishes the edema.

Elevation of the leg, immediate application of a firm and evenly distributed bandage, and immediate application of ice are also advocated. If a sphygmomanometer is not available, or if there is no one present with knowledge of its use, elevation of the leg, ice packing, and application of a pressure dressing are indicated.

A crepe or elastic bandage is preferable to adhesive tape to permit application and reapplication with minimal skin irritation. The bandaging must be firmly and uniformly applied to include the entire foot proximal to the tear as well as the lower half of the leg.

Ice placed within a plastic bag is applied, or preferably, is applied by emerging the entire foot in a container of water and ice cubes. If the leg is occluded by a sphygmomanometer, emergence in ice water is permitted. Ice should be applied at 20- to 30-minute intervals every 2 to 3 hours for the first day. Often, use of cuff occlusion for half an hour to one hour makes further ice application unnecessary.

When walking is resumed, the ankle that has been severely sprained should be bandaged. Use of crutches to prevent or minimize weight bearing is often desirable.

Roentgenographic studies should be performed as soon as possible. If effusion is marked in the presence of a fracture, plaster casting may be delayed. This delay in casting is acceptable if all aspects of immediate treatment are strictly observed.

If the sprain is minor, roentgenographic results are negative, and excessive mobility of the ankle is not evident. The foot and ankle may be rebandaged daily, and ice application continued, for several days.

After 3 to 4 days, heat application may replace ice application. The bandage should *not* be removed too soon, as swelling may return or the ligaments may be resubjected to added stress. Seven to 10 days of bandaging is usually adequate.

Active range-of-motion exercises performed by the patient help disperse edema and prevent adhesions. Range of motion for dorsiflexion, plantar flexion, toe flexion, and gentle inversion and eversion should be included. Competitive sports should be delayed for 1 to 3 weeks, depending on the severity of the sprain.

When a severe lateral ligamentous tear has occurred, the foot should be casted in slight *eversion* and at 90 degrees of neutral dorsiflexion. If edema is present when the cast is applied, the cast can be removed every day or every few days and reapplied. Movement of the foot in the cast indicates the frequency of recasting. In severe sprains, the cast with a walking heel should be continued for approximately 8 to 10 weeks.

Frequently after a severe ankle sprain, the patient complains of instability or a weak ankle that causes insecurity and clumsiness. Balance and coordination are impaired. During examination, poor balance is observed when the patient stands on the injured leg, as compared with the normal leg. This impaired one-foot balance is observed whether the patient's eyes are open or closed, indicating impairment of proprioception. As the ligaments are highly innervated, it is plausible that the sensory end-organs of the nerves within the ligaments can be damaged by stretching during the ligamentous injury. Recovery is apparently incomplete.

Restoration or improvement of balance can be initiated by balancing exercises. Mere practice of one-leg stance and balance is beneficial. By standing on a board that is 4 × 12 × 1/2 inches, and balanced on a half round, the patient can enhance the balance exercise (Fig. 11–48). At first, the half round is placed at a right angle to the length of the board. Balance in this position involves ankle dorsiplantar movement. Gradually, the half

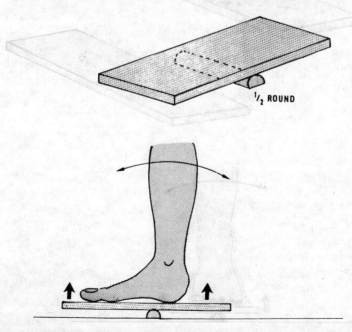

1/2 ROUND

Figure 11–48. Balance exercise. By standing on a 4 × 12 × 1/2-inch board that is balanced on a half round placed at a right angle to its length, the patient can develop balance of ankle dorsiflexion and plantar flexion. This exercise can be performed for 5 minutes several times daily.

round is turned until it is longitudinal to the board (Fig. 11–49). At this point, ankle eversion and inversion are stressed. This treatment gradually improves balance and enhances the ligamentous proprioception.

Surgical repair of severely torn ligaments is usually not necessary if proper casting is applied. If instability remains after proper care is administered, however, or if dislocation recurs, surgical intervention should be considered.

In the injury that tears the medial ligament and the tibiofibular ligaments and widens the ankle mortice, the plaster cast should be applied with the intent of compressing the ankle bilaterally. Reapplication of casts is indicated until the final cast is snug. Weight bearing, even in the cast, should be avoided for a minimum of 8 weeks.

Open reduction of this injury may be necessary, but should be considered only if fragmentation occurs, or if casting fails to achieve reduction.

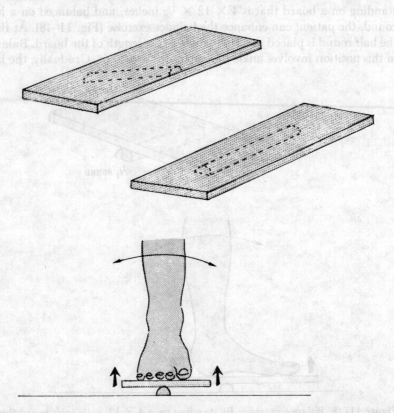

Figure 11–49. Balance exercise. Gradually, the half round under the balance board is rotated until it is placed longitudinally to the board. This change in angulation of the half round requires more intricate balance of the ankle. The half round can be rotated at weekly intervals.

STRESS FRACTURES

With the current increase in avocational sports activities and health-improvement exercise, there has been a marked increase in the so-called "overuse syndromes." These syndromes involve injuries to the lower extremity that are caused by excessive prolonged athletic activities and are often incurred by nonathletes.

Stress fractures are increasingly prevalent. They usually affect the fibula, tibia, calcaneus, and the metatarsals. These fractures occur in sports such as running or jogging—activities in which repetitive mechanical force is imposed upon the long bones of the lower extremity. Often, these injuries occur when there has been a significant increase in mileage or when a new activity has been initiated.

Symptoms are usually "deep pain" and tenderness over the area of fracture. Pain occurs upon resumption of the injurious activity. The diagnosis may be confirmed initially by a bone scan and ultimately by x-ray studies.

Treatment consists essentially of "rest" of the part and local application of ice. As this injury frequently occurs in avid athletes who will not willingly discontinue the running activity, a splint, a constrictive soft shoe with proper orthosis for the fractured foot, or a firm but soft-padded plastic brace for the lower leg is used in the resumption of the activity. Running *in deep water* has been claimed to buoy the athlete, decrease the gravity effect, and maintain the athlete's physical and psychologic status.

REFERENCES

1. Cailliet, R: Foot and Ankle Pain, ed 2. FA Davis, Philadelphia, 1983.
2. Thompson, TC, and Doherty, JH: Spontaneous rupture of tendon of Achilles: A new diagnostic clinical test. J Trauma 2:126, 1962.
3. Dorland's Illustrated Medical Dictionary, ed 26. WB Saunders, Philadelphia, 1981.

BIBLIOGRAPHY

Basmajian, JV: Man's posture. Arch Phys Med 46:26, 1965.
Basmajian, JV, and Stecko, G: The roles of muscles in arch support of the foot. J Bone Joint Surg [Am] 45:1184, 1963.
Brantingham, CR, Egge, AS, and Beekman, BE: The Effect of Artificially Varied Surface on Ambulatory Rehabilitation with Preliminary EMG Evaluation of Certain Muscles Involved. Presented at APA Annual Meeting, Los Angeles, August 1963.
Chomeley, JA: Hallux valgus in adolescents. Proc Roy Soc Med 51:903, 1958.
Clayton, ML, and Weir, GJ: Experimental investigations of ligamentous healing. Am J Surg 98:373, 1959.

DePalma, AF: Section I, Symposium: Injuries to the ankle joint. Clin Orthop Related Res 42:2, 1965.

Du Vries, HL: Surgery of the Foot. CV Mosby, St. Louis, 1965.

Freeman, MAR, Dean, MRE, and Hanham, IWF: The etiology and prevention of functional instability of the foot. J Bone Joint Surg [Br] 47:678, 1965.

Grant, JCB: A Method of Anatomy, ed 5. Williams & Wilkins, Baltimore, 1952.

Griffiths, JC: Tendon injuries around the ankle. J Bone Joint Surg [Br] 47:686, 1965.

Haymaker, W, and Woodhall, B: Peripheral Nerve Injuries: Principles of Diagnosis, ed 2. WB Saunders, Philadelphia, 1953.

Hicks, JH: Mechanics of the foot. J Anat 87:345, 1953.

Hicks, JH: Axis rotation at ankle joint. J Anat 86:1, 1952.

Hollinshead, WH: Functional Anatomy of the Limbs and Back. WB Saunders, Philadelphia, 1952.

Jones, FW: Structures and Function as Seen in the Foot, ed 2. Bailliere, Tindall and Cox, London, 1949.

Jones, FW: Talocalcaneal articulation. Lancet 2:241, 1944.

Kaplan, E: Some principles of anatomy and kinesiology in stabilizing operations of the foot. Clin Orthop 34:7, 1964.

Kelikian, H: Hallux Valgus and Allied Deformities of the Forefoot and Metatarsalgia. WB Saunders, Philadelphia, 1965.

Keller, WL: Surgical treatment of bunions and hallux valgus. NY Med J 80:741, 1904.

Lake, NC: The Foot, ed 3. Bailliere, Tindall and Cox, London, 1943.

Lapidus, PW: Kinesiology and mechanical anatomy of the tarsal joints. Clin Orthop 30:20, 1963.

Lapidus, PW: Operation for correction of hammer toe. J Bone Joint Surg 21:4, 1939.

Lea, RB, and Smith L: Non-surgical treatment of tendo achilles rupture. J Bone Joint Surg [Am] 54:1398–1407, 1972.

Leach, RE: Achilles Tendon Rupture. Am Acad Orthop Surg: Symposium on Foot and Leg and Running Sports. CV Mosby, St. Louis, 1980. Chapter 12, pp 99–103.

Levens, AS, Inman, VT, and Blosser, JA: Transverse rotation of the segments of the lower extremity in locomotion. J Bone Joint Surg [Am] 30:849, 1948.

Lewin, P: The Foot and Ankle, ed 4. Lea & Febiger, Philadelphia, 1959.

Liberson, WT: Biomechanics of gait: A method of study. Arch Phys Med 46:37, 1965.

MacConaill, MA: The postural mechanism of the human foot. Proc R Ir Acad 1-B:265, 1945.

Mann, R, and Inman, VT: Phasic activity of intrinsic muscles of the foot. J Bone Joint Surg [Am] 46:469, 1964.

McBride, ED: A conservative operation for bunions. J Bone Joint Surg 10:735, 1928.

McLaughton, HL: Trauma. WB Saunders, Philadelphia, 1959.

Mennell, J: The Science and Art of Joint Manipulation, ed 2. J & A Churchill, London, 1949.

Meyerding, H, and Shellito, JG: Dupuytren's contracture of the foot. J Int Coll Surg 11:596, 1948.

Milgram, JE: Office procedures for relief of the painful foot. J Bone Joint Surg [Am] 46:1095, 1964.

Morton, DJ: Human Locomotion and Body Form: A Study of Gravity and Man. Williams & Wilkins, Baltimore, 1952.

Moseley, HF: Traumatic disorders of the ankle and foot. Clin Symp 17:3–30, 1965.

Murray, MP, Drought, AB, and Kory, RC: Walking patterns of normal men. J Bone Joint Surg [Am] 46:335, 1964.

Rubin, G, and Witten, M: The talar tilt angle and the fibular collateral ligament. J Bone Joint Surg [Am] 42:311, 1960.

Ryder, CT, and Crane, L: Measuring femoral anteversion: The problem and a method. J Bone Joint Surg [Am] 35:321, 1953.

Saunders, JB de CM, Inman, VT, and Eberhart, HD: The major determinants in normal and pathological gait. J Bone Joint Surg [Am] 35:543, 1953.

Schwartz, RP, and Heath, AL: Pointed and round-toed shoes. J Bone Joint Surg [Am] 48:2, 1966.

Sutherland, DH: An electromyographic study of the plantar flexors of the ankle in normal walking on the level. J Bone Joint Surg [Am] 48:66, 1966.

Taylor, RG: The treatment of claw toes by multiple transfers of flexor into extensor tendons. J Bone Joint Surg [Br] 33:539, 1951.

Thompson, TC, and Doherty, JH: Spontaneous rupture of tendon of Achilles: A new diagnostic clinical test. J Trauma, 2:126, 1962.

Wilson, JN: The treatment of deformities of the foot and toes. Br J Phys Med 17:73, 1954.

Wilson, JN: V-Y correction for varus deformity of the fifth toe. Br J Surg 41:133, 1953.

Wright, DG, Desaid, SM, and Henderson, WH: Action of the Subtalar and Ankle Joint Complex During the Stance Phase of Walking. Biomechanics Laboratory, University of California, San Francisco, No. 38, June 1962.

Zamosky, I, and Licht, S: Shoes and their modifications. In Licht, S (Ed): Orthotics Etcetera. Phys Med Library, Vol 9. Elizabeth Licht, New Haven, 1966.

Index

Page numbers followed by F indicate figures; page numbers followed by T indicate tables.

367